Audio E

Audio Electronics

Second edition

John Linsley Hood

Newnes

OXFORD AUCKLAND BOSTON JOHANNESBURG MELBOURNE NEW DELHI

Newnes
An imprint of Butterworth-Heinemann
Linacre House, Jordan Hill, Oxford OX2 8DP
225 Wildwood Avenue, Woburn, MA 01801-2041
A division of Reed Educational and Professional Publishing Ltd

A member of the Reed Elsevier plc group

First published 1995
Second edition 1999
© Chapters 1–5 Butterworth-Heinemann 1993, 1995, 1998
    Chapters 6–11 John Linsley Hood 1995, 1999

British Library Cataloguing in Publication Data
A catalogue record for this book is available from the British Library

ISBN 0 7506 4332 3

Library of Congress Cataloguing in Publication Data
A catalogue record for this book is available from the Library of Congress

Composition by Genesis Typesetting, Rochester, Kent
Printed and bound in Great Britain by Biddles Ltd, Guildford and King's Lynn

Contents

Preface

The progress of 'audio' towards the still-distant goal of a perfect imitation of reality has been, and remains, inextricably bound up with the development of electronic components and circuit technology, and with the parallel progress in the various transducers and interface devices used to generate and reproduce electrical signals. However, while in the early years of audio almost all the development work was done in an empirical manner, with ideas being tested experimentally in the studios or listening rooms of those involved – for want of any better way of advancing the design technology – gradually, as our understanding of the technical problems and their solutions increased, the way in which the designers make their designs has become increasingly analytical and theoretical in its nature.

This change is as inevitable as it is predictable since there are many parts of this work – digital audio, for example – which can no longer be designed by any pragmatic 'suck it and see' method, and whose undoubted success has been entirely dependent on the correct outcome of theoretical calculations and predictions. Unfortunately, this has left a large number of music and hi-fi enthusiasts in the dark about what is actually being done to achieve the results they hear; and the occasional design errors made by the engineers, which have led to deficiencies in the reproduced sound, have left many listeners suspicious of what they no longer understand.

Design errors still do occur, just as they have always done, but now that the design and marketing decisions are no longer based on a judgement of sound quality made by a knowledgeable enthusiast/designer, there is a greater risk that equipment embodying them will find its way on to the dealers' shelves.

The various hi-fi magazines perform a useful task – irritating though they may be to those who already know everything – in drawing the attention of the engineers to the not entirely infrequent differences between what the specification implies and what the ear actually hears. However, the main requirement for the listener must remain a greater understanding of what is actually done, and how this will influence what he or she hears. This book is an attempt, in one small corner of this field, to reduce this gap between hearing and understanding.

John Linsley Hood

CHAPTER 1

Tape recording

THE BASIC SYSTEM

In principle, the recording of an alternating electrical signal as a series of magnetic fluctuations on a continuous magnetisable tape would not appear to be a difficult matter, since it could be done by causing the AC signal to generate corresponding changes in the magnetic flux across the gap of an electromagnet, and these could then be impressed on the tape as it passes over the recording electromagnet head.

In practice, however, there are a number of problems, and the success of tape recording, as a technique, depends upon the solution of these, or, at least, on the attainment of some reasonable working compromise. The difficulties which exist, and the methods by which these are overcome, where possible, are considered here in respect of the various components of the system.

MAGNETIC TAPE

This is a thin continuous strip of some durable plastics base material, which is given a uniform coating of a magnetisable material, usually either 'gamma' ferric oxide (Fe_2O_3), chromium dioxide (CrO_2), or, in some recently introduced tapes, of a metallic alloy, normally in powder form, and held by some suitable binder material. Various 'dopants' can also be added to the coating, such as cobalt, in the case of ferric oxide tapes, to improve the magnetic characteristics.

To obtain a long playing time it is necessary that the total thickness of the tape shall be as small as practicable, but to avoid frequency distortion on playback it is essential that the tape shall not stretch in use. It is also important that the surface of the tape backing material shall be hard, smooth and free from lumps of imperfectly extruded material (known as 'pollywogs') to prevent inadvertent momentary loss of contact between the tape and the recording or play-back heads, which would cause 'dropouts' (brief interruptions in the replayed signal). The tape backing material should also be unaffected, so far as is possible, by changes in temperature or relative humidity.

For cassette tapes, and other systems where a backup pressure pad is

1

used, the uncoated surface is chosen to have a high gloss. In other applications a matt finish will be preferred for improved spooling.

The material normally preferred for this purpose, as the best compromise between cost and mechanical characteristics, is biaxially oriented polyethylene terephthalate film (Melinex, Mylar, or Terphan). Other materials may be used as improvements in plastics technology alter the cost/performance balance.

The term 'biaxial orientation' implies that these materials will be stretched in both the length and width directions during manufacture, to increase the surface smoothness (gloss), stiffness and dimensional stability (freedom from stretch). They will normally also be surface treated on the side to which the coating is to be applied, by an electrical 'corona discharge' process, to improve the adhesion of the oxide containing layer. This is because it is vitally important that the layer is not shed during use as it would contaminate the surface or clog up the gaps in the recorder heads, or could get into the mechanical moving parts of the recorder.

In the tape coating process the magnetic material is applied in the form of a dope, containing also a binder, a solvent and a lubricant, to give an accurately controlled coating thickness. The coated surface is subsequently polished to improve tape/head contact and lessen head wear. The preferred form of both ferric oxide and chromium dioxide crystals is needle-shaped, or 'acicular', and the best characteristic for audio tapes are given when these are aligned parallel to the surface, in the direction of magnetisation. This is accomplished during manufacture by passing the tape through a strong, unidirectional magnetic field, before the coating becomes fully dry. This aligns the needles in the longitudinal direction. The tape is then demagnetised again before sale.

Chromium dioxide and metal tapes both have superior properties, particularly in HF performance, resistance to 'print through' and deterioration during repeated playings, but they are more costly. They also require higher magnetic flux levels during recording and for bias and erase purposes, and so may not be suitable for all machines.

The extra cost of these tape formulations is normally only considered justifiable in cassette recorder systems, where reproduction of frequencies in the range 15–20 kHz, especially at higher signal levels, can present difficulties.

During the period in which patent restrictions limited the availability of chromium dioxide tape coatings, some of the manufacturers who were unable to employ these formulations for commercial reasons, put about the story that chromium dioxide tapes were more abrasive than iron oxide ones. They would, therefore, cause more rapid head wear. This was only marginally true, and now that chromium dioxide formulations are more widely available, these are used by most manufacturers for their premium quality cassette tapes.

Table 1.1 Tape thicknesses (reel-to-reel)

Tape	Thickness (in.)
'Standard play'	0.002
'Long play'	0.0015
'Double play'	0.001
'Triple play'	0.00075
'Quadruple play'	0.0005

Composite 'ferro-chrome' tapes, in which a thinner surface layer of a chromium dioxide formulation is applied on top of a base ferric oxide layer, have been made to achieve improved HF performance, but without a large increase in cost.

In 'reel-to-reel' recorders, it is conventional to relate the tape thickness to the relative playing time, as 'Standard Play', 'Double Play' and so on. The gauge of such tapes is shown in Table 1.1. In cassette tapes, a more straightforward system is employed, in which the total playing time in minutes is used, at the standard cassette playing speed. For example, a C60 tape would allow 30 minutes playing time, on each side. The total thicknesses of these tapes are listed in Table 1.2.

For economy in manufacture, tape is normally coated in widths of up to 48 in. (1.2 m), and is then slit down to the widths in which it is used. These are 2 in. (50.8 mm), 1 in. (25.4 mm), 0.5 in. (12.7 mm) and 0.25 in. (6.35 mm) for professional uses, and 0.25 in. for domestic reel-to-reel machines. Cassette recorders employ 0.15 in. (3.81 mm) tape.

High-speed slitting machines are complex pieces of precision machinery which must be maintained in good order if the slit tapes are to have the required parallelism and constancy of width. This is particularly important in cassette machines where variations in tape width can cause bad winding, creasing, and misalignment over the heads.

Table 1.2 Tape thicknesses (cassette)

Tape	Thickness (μm)
C60	18 (length 92 m)
C90	12 (length 133 m)
C120	9 (length 184 m)

Tape base thicknesses 12 μm, 8 μm and 6 μm respectively.

For all of these reasons, it is highly desirable to employ only those tapes made by reputable manufacturers, where these are to be used on good recording equipment, or where permanence of the recording is important.

THE RECORDING PROCESS

The magnetic materials employed in tape coatings are chosen because they possess elemental permanent magnets on a sub-microscopic or molecular scale. These tiny magnetic elements, known as 'domains', are very much smaller than the grains of spherical or needle-shaped crystalline material from which oxide coatings are made.

Care will be taken in the manufacture of the tape to try to ensure that all of these domains will be randomly oriented, with as little 'clumping' as possible, to obtain as low a zero-signal-level noise background as practicable. Then, when the tape passes over a recording head, shown schematically in Fig. 1.1, these magnetic domains will be realigned in a direction and to an extent which depend on the magnetic polarity and field strength at the trailing edge of the recording head gap.

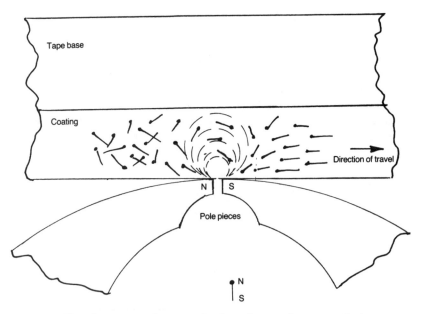

Fig. 1.1   *The alignment of magnetic domains as the magnetic tape passes over the recording head.*

This is where first major snag of the system appears. Because of the magnetic inertia of the material, small applied magnetic fields at the recording head will have very little effect in changing the orientation of the domains. This leads to the kind of characteristic shown in Fig. 1.2, where the applied magnetising force, (*H*), is related to the induced flux density in the tape material, (*B*).

If a sinusoidal signal is applied to the head, and the flux across the recording head gap is related to the signal voltage, as shown in Fig. 1.2, the remanent magnetic flux induced in the tape – and the consequent replayed signal – would be both small in amplitude and badly distorted.

This problem is removed by applying a large high-frequency signal to the recording head, simultaneously with the desired signal. This superimposed HF signal is referred to as 'HF bias' or simply as 'bias', and

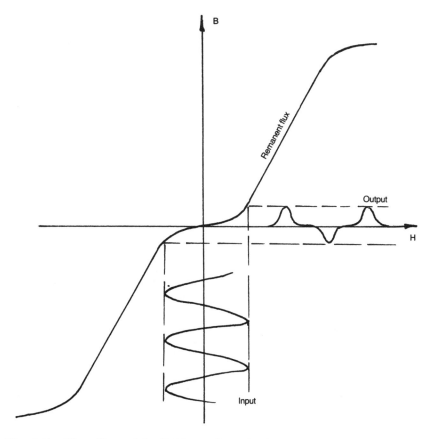

Fig. 1.2 *The effect of the B-H non-linearity in magnetic materials on the recording process.*

will be large enough to overcome the magnetic inertia of the domains and take the operating region into the linear portion of the BH curve.

Several theories have been offered to account for the way in which 'HF bias' linearises the recording process. Of these the most probable is that the whole composite signal is in fact recorded but that the very high frequency part of it decays rapidly, due to self cancellation, so that only the desired signal will be left on the tape, as shown in Fig. 1.3.

When the tape is passed over the replay head — which will often be the same head which was used for recording the signal in the first place — the

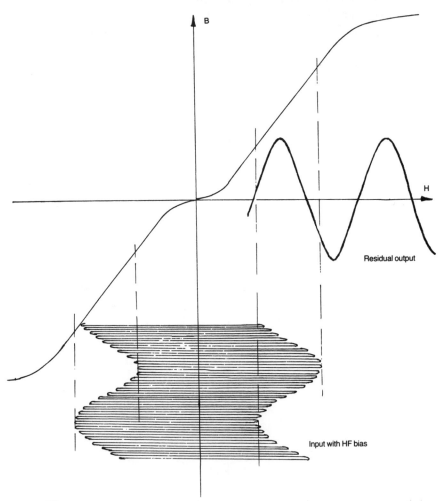

Fig. 1.3   *The linearising effect of superimposed HF bias on the recording process.*

fluctuating magnetic flux of the tape will induce a small voltage in the windings of the replay head, which will then be amplified electronically in the recorder. However, both the recorded signal, and the signal recovered from the tape at the replay head will have a non-uniform frequency response, which will demand some form of response equalisation in the replay amplifier.

CAUSES OF NON-UNIFORM FREQUENCY RESPONSE

If a sinusoidal AC signal of constant amplitude and frequency is applied to the recording head and the tape is passed over this at a constant speed, a sequence of magnetic poles will be laid down on the tape, as shown schematically in Fig. 1.4, and the distance between like poles, equivalent to one cycle of recorded signal, is known as the recorded wavelength. The length of this can be calculated from the applied frequency and the movement, in unit time of the tape, as 'λ', which is equal to tape velocity (cm/s) divided by frequency (cycles/s) which will give wavelength (λ) in cm.

This has a dominant influence on the replay characteristics in that when the recorded wavelength is equal to the replay head gap, as is the case shown at the replay head in Fig. 1.4, there will be zero output. This situation is worsened by the fact that the effective replay head gap is, in practice, somewhat larger than the physical separation between the opposed pole pieces, due to the spread of the magnetic field at the gap, and to the distance between the head and the centre of the magnetic coating on the tape, where much of the HF signal may lie.

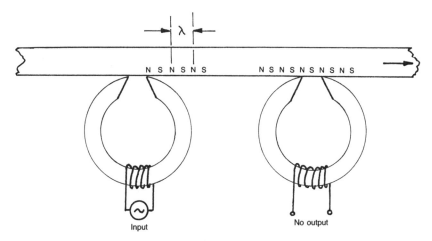

Fig. 1.4 *The effect of recorded wavelength on replay head output.*

Additional sources of diminished HF response in the recovered signal
are eddy-current and other magnetic losses in both the recording and
replay heads, and 'self-demagnetisation' within the tape, which increases
as the separation between adjacent N and S poles decreases.

If a constant amplitude sine-wave signal is recorded on the tape, the
combined effect of recording head and tape losses will lead to a remanent

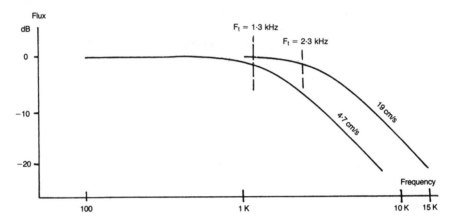

Fig. 1.5   *Effect of tape speed on HF replay output.*

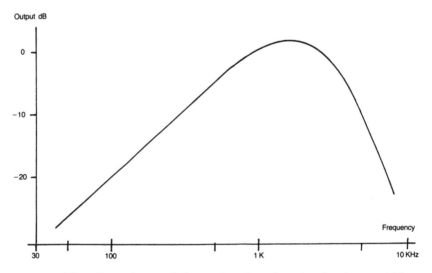

Fig. 1.6   *The effect of recorded wavelength and replay head gap width on
replay signal voltage.*

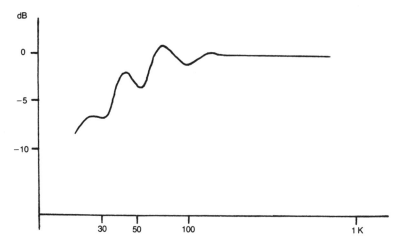

Fig. 1.7 *Non-uniformity in low-frequency response due to pole-piece contour effects.*

flux on the tape which is of the form shown in Fig. 1.5. Here the HF turnover frequency (F_t) is principally dependent on tape speed.

Ignoring the effect of tape and record/replay head losses, if a tape, on which a varying frequency signal has been recorded, is replayed at a constant linear tape speed, the signal output will increase at a linear rate with frequency. This is such that the output will double for each octave of frequency. The result is due to the physical laws of magnetic induction, in which the output voltage from any coil depends on the magnetic flux passing through it, according to the relationship $V = L.\mathrm{d}B/\mathrm{d}t$. Combining this effect with replay head losses leads to the kind of replay output voltage characteristics shown in Fig. 1.6.

At very low frequencies, say below 50 Hz, the recording process becomes inefficient, especially at low tape speeds. The interaction between the tape and the profile of the pole faces of the record/replay heads leads to a characteristic undulation in the frequency response of the type shown in Fig. 1.7 for a good quality cassette recorder.

RECORD/REPLAY EQUALISATION

In order to obtain a flat frequency response in the record/replay process, the electrical characteristics of the record and replay amplifiers are modified to compensate for the non-linearities of the recording process, so far as this is possible. This electronic frequency response adjustment is known as 'equalisation', and is the subject of much misunderstanding, even by professional users.

This misunderstanding arises because the equalisation technique is of a different nature to that which is used in FM broadcast receivers, or in the reproduction of LP gramophone records, where the replay de-emphasis at HF is identical in time-constant (turnover frequency) and in magnitude, but opposite in sense to the pre-emphasis employed in transmission or recording.

By contrast, in tape recording it is assumed that the total inadequacies of the recording process will lead to a remanent magnetic flux in the tape following a recording which has been made at a constant amplitude which has the frequency response characteristics shown in Fig. 1.8, for various tape speeds. This remanent flux will be as specified by the time-constants quoted for the various international standards shown in Table 1.3.

These mainly refer to the expected HF roll-off, but in some cases also require a small amount of bass pre-emphasis so that the replay de-emphasis – either electronically introduced, or inherent in the head response – may lessen 'hum' pick-up, and improve LF signal-to-noise (S/N) ratio.

The design of the replay amplifier must then be chosen so that the required flat frequency response output would be obtained on replaying a tape having the flux characteristics shown in Fig. 1.8.

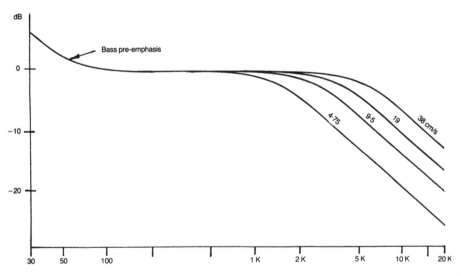

Fig. 1.8 *Assumed remanent magnetic flux on tape for various tape speeds, in cm/s.*

This will usually lead to a replay characteristic of the kind shown in Fig. 1.9. Here a −6 dB/octave fall in output, with increasing frequency, is needed to compensate for the increasing output during replay of a constant recorded signal − referred to above − and the levelling off, shown in curves a–d, is simply that which is needed to correct for the anticipated fall in magnetic flux density above the turn-over frequency, shown in Fig. 1.8 for various tape speeds.

However, this does not allow for the various other head losses, so some additional replay HF boost, as shown in the curves e–h, of Fig. 1.9, is also used.

The recording amplifier is then designed in the light of the performance of the recording head used, so that the remanent flux on the tape will conform to the specifications shown in Table 1.3, and as illustrated in Fig. 1.8. This will generally also require some HF (and LF) pre-emphasis, of the kind shown in Fig. 1.10. However, it should be remembered that, especially in the case of recording amplifier circuitry the componet values. chosen by the designer will be appropriate only to the type of recording head and tape transport mechanism − in so far as this may influence the tape/head contact − which is used in the design described.

Because the subjective noise level of the system is greatly influenced by the amount of replay HF boost which is employed, systems such as reel-to-reel recorders, operating at relatively high-tape speeds, will sound less 'hissy' than cassette systems operating at 1.875 in./s (4.76 cm/s), for which substantial replay HF boost is necessary. Similarly, recordings made using

Table 1.3 Frequency correction standards

| Tape speed | Standard | Time constants (µs) | | −3 dB@ | +3 dB@ |
		HF	LF	HF (kHz)	LF (Hz)
15 in./s	NAB	50	3180	3.18	50
(38.1 cm/s)	IEC	35	—	4.55	—
	DIN	35	—	4.55	—
7½ in./s	NAB	50	3180	3.18	50
(19.05 cm/s)	IEC	70	—	2.27	—
	DIN	70	—	2.27	—
3¾ in./s	NAB	90	3180	1.77	50
(9.53 cm/s)	IEC	90	3180	1.77	50
	DIN	90	3180	1.77	50
1⅞ in./s	DIN (Fe)	120	3180	1.33	50
(4.76 cm/s)	(CrO$_2$)	70	3180	2.27	50

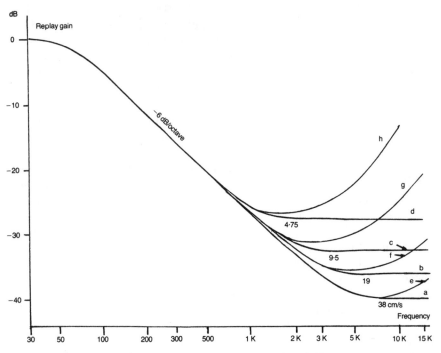

Fig. 1.9 *Required replay frequency response, for different tape speeds.*

Chromium dioxide tapes, for which a 70 μs HF emphasis is employed, will sound less 'hissy' than with Ferric oxide tapes equalised at 120 μs, where the HF boost begins at a lower frequency.

HEAD DESIGN

Record/replay heads

Three separate head types are employed in a conventional tape recorder for recording, replaying and erasing. In some cases, such as the less expensive reel-to-reel and cassette recorders, the record and replay functions will be combined in a single head, for which some design compromise will be sought between the different functions and design requirements.

In all cases, the basic structure is that of a ring electromagnet, with a flattened surface in which a gap has been cut, at the position of contact with the tape. Conventionally, the form of the record and replay heads is as shown in Fig. 1.11, with windings symmetrically placed on the two limbs, and with gaps at both the front (tape side) and the rear. In both

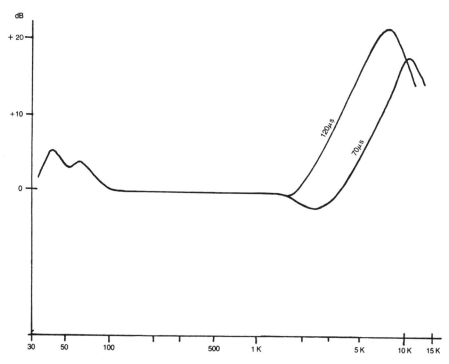

Fig. 1.10 *Probable replay frequency response in cassette recorder required to meet flux characteristics shown in Fig. 1.8.*

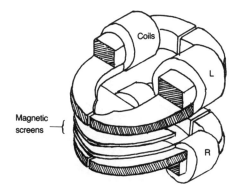

Fig. 1.11 *General arrangement of cassette recorder stereo record/replay head using Ferrite pole pieces.*

cases the gaps will be filled with a thin shim of low magnetic permeability material, such as gold, copper or phosphor bronze, to maintain the accuracy of the gap and the parallelism of the faces.

The pole pieces on either side of the front gap are shaped, in conjunction with the gap filling material, to concentrate the magnetic field in the tape, as shown schematically in Fig. 1.12, for a replay head, and the material from which the heads are made is chosen to have as high a value of permeability as possible, for materials having adequate wear resistance.

The reason for the choice of high permeability core material is to obtain as high a flux density at the gap as possible when the head is used for recording, or to obtain as high an output signal level as practicable (and as high a resultant S/N ratio) when the head is used in the replay mode. High permeability core materials also help to confine the magnetic flux within the core, and thereby reduce crosstalk.

With most available ferromagnetic materials, the permeability decreases with frequency, though some ferrites (sintered mixes of metallic oxides) may show the converse effect. The rear gap in the recording heads is used to minimise this defect. It also makes the recording efficiency less dependent on intimacy of the tape to head contact. In three-head machines, the rear gap in the replay head is often dispensed with, in the interests of the best replay sensitivity.

Practical record/replay heads will be required to provide a multiplicity of signal channels on the tape: two or four, in the case of domestic systems, or up to 24 in the case of professional equipment. So a group of identical heads will require to be stacked, one above the other and with

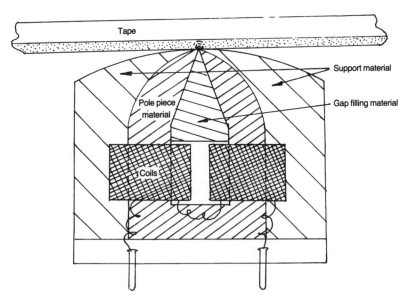

Fig. 1.12 *Relationship between pole-pieces and magnetic flux in tape.*

the gaps accurately aligned vertically to obtain accurate time coincidence of the recorded or replayed signals.

It is, of course, highly desirable that there should be very little interaction, or crosstalk, between these adjacent heads, and that the recording magnetic field should be closely confined to the required tape track. This demands very careful head design, and the separation of adjacent heads in the stack with shims of low permeability material.

To preserve the smoothness of the head surface in contact with the tape, the non-magnetic material with which the head gap, or the space between head stacks, is filled is chosen to have a hardness and wear resistance which matches that of the head alloy. For example, in Permalloy or Super Permalloy heads, beryllium copper or phosphor bronze may be used, while in Sendust or ferrite heads, glass may be employed.

The characteristics of the various common head materials are listed in Table 1.4. Permalloy is a nickel, iron, molybdenum alloy, made by a manufacturing process which leads to a very high permeability and a very low coercivity. This term refers to the force with which the material resists demagnetisation: 'soft' magnetic materials have a very low coercivity, whereas 'hard' magnetic materials, i.e. those used for 'permanent' magnets, have a very high value of coercivity.

Super-Permalloy is a material of similar composition which has been heat treated at 1200−1300° C in hydrogen, to improve its magnetic properties and its resistance to wear. Ferrites are ceramic materials, sintered at a high temperature, composed of the oxides of iron, zinc, nickel and manganese, with suitable fillers and binders. As with the metallic alloys, heat treatment can improve the performance, and hot-pressed ferrite (typically 500 kg/cm^2 at 1400° C) offers both superior hardness and better magnetic properties.

Sendust is a hot-pressed composite of otherwise incompatible metallic alloys produced in powder form. It has a permeability comparable to that of Super Permalloy with a hardness comparable to that of ferrite.

In the metallic alloys, as compared with the ferrites, the high conductivity of the material will lead to eddy-current losses (due to the core material behaving like a short-circuited turn of winding) unless the material is made in the form of thin laminations, with an insulating layer between these. Improved HF performance requires that these laminations are very thin, but this, in turn, leads to an increased manufacturing cost, especially since any working of the material may spoil its performance. This would necessitate re-annealing, and greater problems in attaining accurate vertical alignment of the pole piece faces in the record/replay heads. The only general rule is that the better materials will be more expensive to make, and more difficult to fabricate into heads.

Because it is only the trailing edge of the record head which generates the remanent flux on the tape, the gap in this head can be quite wide; up

Table 1.4 Magnetic materials for recording heads

Material	Mumetal	Permalloy	Super Permalloy	Sendust (Hot pressed)	Ferrite	Hot Pressed Ferrite
Composition	75 Ni 2 Cr 5 Cu 18 Fe	79 Ni 4 Mo 17 Fe	79 Ni 5 Mo 16 Fe	85 Fe 10 Si 5 Al	Mn Ni Fe oxides Zn	Mn Ni Fe oxides Zn
Treatment	1100° C in hydrogen	1100° C	1300° C in hydrogen	800° C in hydrogen		500 kg/cm^2 in hydrogen
Permeability 1 kHz	50 000	25 000	200 000*	50 000*	1200	20 000
Max flux density (gauss)	7200	16 000	8700	> 5000*	4000	4000
Coercivity (oersteds)	0.03	0.05	0.004	0.03	0.5	0.015
Conductivity	High	High	High	High	Very low	Very low
Vickers hardness	118	132	200*	280*	400	700

*Depends on manufacturing process.

to 6–10 μm typical examples. On the other hand, the replay head gap should be as small as possible, especially in cassette recorders, where high quality machines may employ gap widths of one μm (0.00004 in.) or less.

The advantage of the wide recording head gap is that it produces a field which will penetrate more fully into the layer of oxide on the surface of the tape, more fully utilising the magnetic characteristics of the tape material. The disadvantage of the narrow replay gap, needed for good HF response, is that it is more difficult for the alternating zones of magnetism on the recorded tape to induce changes in magnetic flux within the core of the replay head. Consequently the replay signal voltage is less, and the difficulty in getting a good S/N ratio will be greater.

Other things being equal, the output voltage from the replay head will be directly proportional to the width of the magnetic track and the tape speed. On both these counts, therefore, the cassette recorder offers an inferior replay signal output to the reel-to-reel system.

The erase head

This is required to generate a very high frequency alternating magnetic flux within the tape coating material. It must be sufficiently intense to take the magnetic material fully into saturation, and so blot out any previously existing signal on the tape within the track or tracks concerned.

The design of the head gap or gaps should allow the alternating flux decay slowly as the tape is drawn away from the gap, so that the residual flux on the tape will fall to zero following this process.

Because of the high frequencies and the high currents involved, eddy-current losses would be a major problem in any laminated metal construction. Erase heads are therefore made from ferrite materials.

A rear air gap is unnecessary in an erase head, since it operates at a constant frequency. Also, because saturating fields are used, small variations in the tape to head contact are less important.

Some modern cassette tape coating compositions have such a high coercivity and remanence (retention of impressed magnetism) that single gap heads may not fully erase the signal within a single pass. So dual gap heads have been employed, particularly on machines offering a 'metal tape' facility. This gives full erasure, (a typical target value of −70 dB being sought for the removal of a fully recorded signal), without excessive erase currents being required, which could lead to overheating of the erase head.

It is important that the erase head should have an erase field which is closely confined to the required recorded tracks, and that it should remain cool. Otherwise unwanted loss of signal or damage to the tape may occur when the tape recording is stopped during the recording process, by the use of the 'pause' control.

In some machines provision is made for switching off the erase head, so that multiple recordings may be overlaid on the same track. This is an imperfect answer to this requirement, however, since every subsequent recording pass will erase pre-existing signals to some extent.

This facility would not be practicable on inexpensive recorders, where the erase head coil is used as the oscillator coil in the erase and 'bias' voltage generator circuit.

RECORDING TRACK DIMENSIONS

These conform to standards laid down by international agreements, or, in the case of the Philips compact cassette, within the original patent specifications, and are as shown in Fig. 1.13.

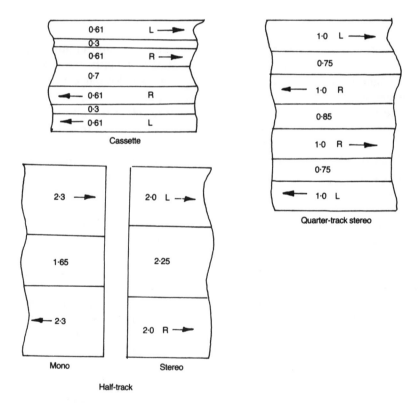

Fig. 1.13 *Tape track specifications. (All dimensions in mm.)*

HF BIAS

Basic bias requirements

As has been seen, the magnetic recording process would lead to a high level of signal distortion, were it not for the fact that a large, constant amplitude. HF bias waveform is combined with the signal at the recording head. The basic requirement for this is to generate a composite magnetic flux, within the tape, which will lie within the linear range of the tape's magnetic characteristics, as shown Fig. 1.3.

However, the magnitude and nature of the bias signal influences almost every other aspect of the recording process. There is no value for the applied 'bias' current which will be the best for all of the affected recording characteristics. It is conventional, and correct, to refer to the bias signal as a current, since the effect on the tape is that of a magnetising force, defined in 'ampere turns'. Since the recording head will have a significant inductance − up to 0.5 H in some cases − which will offer a substantial impedance at HF., the applied voltage for a constant flux level will depend on the chosen bias frequency.

The way in which these characteristics are affected is shown in Figs 1.14 and 1.15, for ferric and chrome tapes. These are examined separately below, together with some of the other factors which influence the final performance.

HF bias frequency

The particular choice of bias frequency adopted by the manufacturer will be a compromise influenced by his own preferences, and his performance intentions for the equipment.

It is necessary that the frequency chosen shall be sufficiently higher than the maximum signal frequency which is to be recorded that any residual bias signal left on the tape will not be reproduced by the replay head or amplifier, where it could cause overload or other undesirable effects.

On the other hand, too high a chosen bias frequency will lead to difficulties in generating an adequate bias current flow through the record head. This may lead to overheating in the erase head.

Within these overall limits, there are other considerations which affect the choice of frequency. At the lower end of the usable frequency range, the overall efficiency of the recording process, and the LF S/N ratio is somewhat improved. The improvement is at the expense of some erasure of the higher audio frequencies, which increases the need for HF pre-emphasis to achieve a 'flat' frequency response. However, since it is

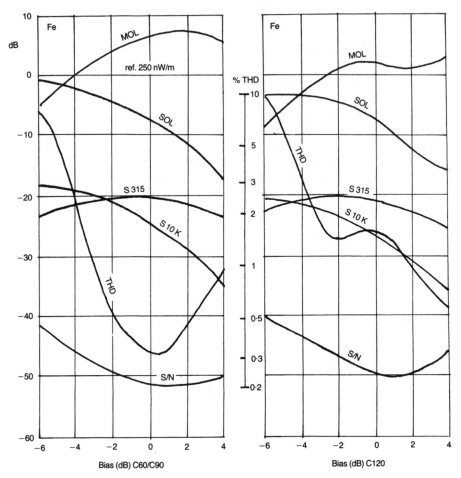

Fig. 1.14 *The influence of bias on recording characteristics. (C60/90 and C120 Ferric tapes.)*

conventional to use the same HF signal for the erase head, the effectiveness of erasure will be better for the same erase current.

At the higher end of the practicable bias frequency range, the recorded HF response will be better, but the modulation noise will be less good. The danger of saturation of the recording head pole-piece tips will be greater, since there is almost invariably a decrease in permeability with increasing frequency. This will require more care in the design of the recording head.

Typically, the chosen frequency for audio work will be between four and six times the highest frequency which it is desired to record: it will usually be in the range 60–120 kHz, with better machines using the higher values.

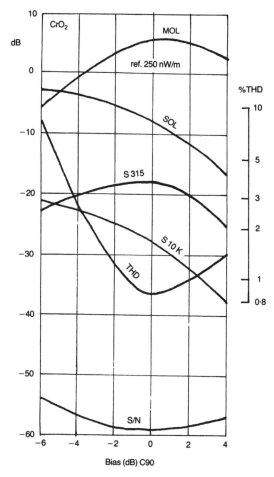

Fig. 1.15 *The influence of bias on recording characteristics. (C90 Chrome tape.)*

HF bias waveform

The most important features of the bias waveform are its peak amplitude and its symmetry. So although other waveforms, such as square waves, will operate in this mode quite satisfactorily, any inadvertent phase shift in the harmonic components would lead to a change in the peak amplitude, which would affect the performance.

Also, it is vitally important that the HF 'bias' signal should have a very low noise component. In the composite signal applied to the record head, the bias signal may be 10–20 times greater than the audio signal to be recorded. If, therefore, the bias waveform is not to degrade the overall

S/N ratio of the system, the noise content of the bias waveform must be at least 40 times better than that of the incoming signal.

The symmetry of the waveform is required so that there is no apparent DC or unidirectional magnetic component of the resultant waveform, which could magnetise the record head and impair the tape modulation noise figure. An additional factor is that a symmetrical waveform allows the greatest utilisation of the linear portion of the tape magnetisation curve.

For these reasons, and for convenience in generating the erase waveform, the bias oscillator will be designed to generate a very high purity sine wave. Then the subsequent handling and amplifying circuitry will be chosen to avoid any degradation of this.

Maximum output level (MOL)

When the tape is recorded with the appropriate bias signal, it will have an effective 'B-H' characteristic of the form shown in Fig. 1.16. Here it is quite linear at low signal levels, but will tend to flatten-off the peaks of the waveform at higher signal levels. This causes the third harmonic distortion, which worsens as the magnitude of the recorded signal increases, and effectively sets a maximum output level for the recorder/tape combination.

This is usually quoted as 'MOL (315 Hz)', and is defined as the output level, at 315 Hz. It is given some reference level, (usually a remanent magnetic flux on the tape of 250 nanoWebers/m, for cassette tapes), at which the third harmonic distortion of the signal reaches 3%. The MOL generally increases with bias up to some maximum value, as shown in Figs 1.14 and 1.15.

The optimum, or reference bias

Since it is apparent that there is no single best value for the bias current, the compromise chosen for cassette recorders is usually that which leads to the least difficulty in obtaining a flat frequency response. Lower values of bias give better HF performance but worse characteristics in almost all other respects, so the value chosen is that which leads to a fall in output, at 10 kHz in the case of cassettes of 12 dB in comparison with the output at 315 Hz.

In higher tape speed reel-to-reel recorders, the specified difference in output level at 315 Hz and 12.5 kHz may be only 2.5 dB. Alternatively, the makers recommended 'reference' bias may be that which gives the lowest value of THD at 1 kHz.

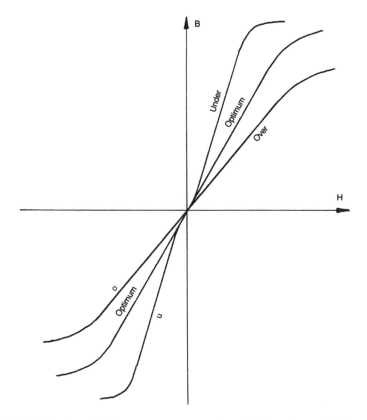

Fig. 1.16 *Relationship between remanent flux ('B'), and applied magnetic field at 315 Hz in a biased tape, showing linearisation and overload effects.*

Sensitivity

This will be specified at some particular frequency, as 'S-315 Hz' or 'S-10 kHz' depending on the frequency chosen. Because of self-erasure effects, the HF components of the signal are attenuated more rapidly with increasing HF 'bias' values than the LF signals. On the other hand, HF signals can act, in part, as a bias waveform, so some recorders employ a monitoring circuit which adjusts the absolute value of the bias input, so that it is reduced in the presence of large HF components in the audio signal.

This can improve the HF response and HF output level for the same value of harmonic distortion in comparison with a fixed bias value system.

Total harmonic distortion (THD)

This is generally worsened at both low and high values of 'bias' current.
At low values, this is because inadequate bias has been used to straighten
out the kink at the centre of the 'B-H' curve, which leaves a residue of
unpleasant, odd harmonic, crossover-type distortion. At high values, the
problem is simply that the bias signal is carrying the composite recording
waveform into regions of tape saturation, where the B-H curve flattens
off.

Under this condition, the distortion component is largely third harmonic,
which tends to make the sound quality rather shrill. As a matter of
practical convenience, the reference bias will be chosen to lie somewhere
near the low point on the THD/bias curve. The formulation and coating
thickness adopted by the manufacturer can be chosen to assist in this, as
can be seen by the comparison between C90 and C120 tapes in Fig. 1.14.

The third harmonic distortion for a correctly biased tape will increase
rapidly as the MOL value is approached. The actual peak recording
values for which the recorder VU or peak recording level meters are set is
not, sadly, a matter on which there is any agreement between manufacturers, and will, in any case, depend on the tape and the way in which it
has been biased.

Noise level

Depending on the tape and record head employed, the residual noise
level on the tape will be influenced by the bias current and will tend to
decrease with increasing bias current values. However, this is influenced
to a greater extent by the bias frequency and waveform employed, and by
the construction of the erase head.

The desired performance in a tape or cassette recorder is that the no-
signal replay noise from a new or bulk erased tape should be identical to
that which arises from a single pass through the recorder, set to record at
zero signal level.

Aspects of recording and replay noise are discussed more fully below.

Saturation output level (SOL)

This is a similar specification to the MOL, but will generally apply to the
maximum replay level from the tape at a high frequency, usually 10 kHz,
and decreases with increasing bias at a similar rate to that of the HF
sensitivity value.

The decrease in HF sensitivity with increasing bias appears to be a simple matter of partial erasure of short recorded wavelength signals by the bias signal. However, in the case of the SOL, the oscillation of the domains caused by the magnetic flux associated with the bias current appears physically to limit the ability of short wavelength magnetic elements to coexist without self-cancellation.

It is fortunate that the energy distribution on most programme material is high at LF and lower middle frequencies, but falls off rapidly above, say, 3 kHz, so HF overload is not normally a problem. It is also normal practice to take the output to the VU or peak signal metering circuit from the output of the recording amplifier, so the effect of pre-emphasis at HF will be taken into account.

Bias level setting

This is a difficult matter in the absence of appropriate test instruments, and will require adjustment for every significant change in tape type, especially in cassette machines, where a wide range of coating formulations is available.

Normally, the equipment manufacturer will provide a switch selectable choice of pre-set bias values, labelled in a cassette recorder, for example, as types '1' (all ferric oxide tapes, used with a 120 μs equalisation time constant), '2' (all chromium dioxide tapes, used with a 70 μs equalisation), '3' (dual coated 'ferro-chrome' tapes, for 70 μs equalisation) and '4' (all 'metal' tapes), where the machine is compatible with these.

Additionally, in the higher quality (three-head) machines, a test facility may be provided, such as a built-in dual frequency oscillator, having outputs at 330 Hz and 8 kHz, to allow the bias current to be set to provide an identical output at both frequencies on replay. In some recent machines this process has been automated.

THE TAPE TRANSPORT MECHANISM

The constancy and precision of the tape speed across the record and replay heads is one of the major criteria for the quality of a tape recorder mechanism. Great care is taken in the drive to the tape 'capstans' to smooth out any vibration or flutter generated by the drive motor or motors. The better quality machines will employ a dual-capstan system, with separate speed or torque controlled motors, to drive the tape and maintain a constant tension across the record and replay heads, in addition to the drive and braking systems applied to the take-up and unwind spools.

This requirement can make some difficulties in the case of cassette recorder systems, in which the design of the cassette body allows only limited access to the tape. In this case a dual-capstan system requires the use of the erase head access port for the unwind side capstan, and forces the use of a narrowed erase head which can be fitted into an adjacent, unused, small slot in the cassette body.

The use of pressure pads to hold the tape against the record or replay heads is considered by many engineers to be an unsatisfactory practice, in that it increases the drag on the tape, and can accelerate head wear. In reel-to-reel recorders it is normally feasible to design the tape layout and drive system so that the tape maintains a constant and controlled position in relation to the heads, so the use of pressure pads can be avoided.

However, in the case of cassettes where the original patent specification did not envisage the quality of performance sought and attained in modern cassette decks, a pressure pad is incorporated as a permanent part of the cassette body. Some three-head cassette machines, with dual capstan drive, control the tape tension at the record and replay position so well that the inbuilt pressure pad is pushed away from the tape when the cassette is inserted.

An essential feature of the maintenance of any tape or cassette recorder is the routine cleaning of the capstan drive shafts, and pinch rollers, since these can attract deposits of tape coating material, which will interfere with the constancy of tape drive speed.

The cleanliness of the record/replay heads influences the closeness of contact between the tape and the head gap, which principally affects the replay characteristics, and reduces the replay HF response.

TRANSIENT PERFORMANCE

The tape recording medium is unique as the maximum rate of change of the reproduced replay signal voltage is determined by the speed with which the magnetised tape is drawn past the replay head gap. This imposes a fixed 'slew-rate' limitation on all reproduced transient signal voltages, so that the maximum rate of change of voltage cannot exceed some fixed proportion of its final value.

In this respect it differs from a slew-rate-limited electronic amplifier, in which the effective limitation is signal amplitude dependent, and may not occur on small signal excursions even when it will arise on larger ones.

Slew-rate limitation in electronic amplifiers is, however, also associated with the paralysis of the amplifier during the rate-limited condition. This is a fault which does not happen in the same way with a replayed magnetic tape signal, within the usable region of the B-H curve. Nevertheless, it is a factor which impairs the reproduced sound quality, and leads to the acoustic superiority of high tape speed reel-to-reel machines over even the best of cassette recorders, and the superiority of replay heads with very narrow head gaps, for example in three-head machines, in comparison with the larger, compromise form, gaps of combined record/ replay heads.

It is possibly also, a factor in the preference of some audiophiles for direct-cut discs, in which tape mastering is not employed.

A further source of distortion on transient waveforms arises due to the use of HF emphasis, in the recording process and in replay, to compensate for the loss of high frequencies due to tape or head characteristics.

If this is injudiciously applied, this HF boost can lead to 'ringing' on a transient, in the manner shown, for a square-wave in Fig. 1.17. This impairs the audible performance of the system, and can worsen the subjective noise figure of the recorder. It is therefore sensible to examine the square-wave performance of a tape recorder system following any adjustment to the HF boost circuitry, and adjust these for minimum overshoot.

TAPE NOISE

Noise in magnetic recording systems has two distinct causes: the tape, and the electronic amplification and signal handling circuitry asso-ciated with it. Circuit noise should normally be the lesser problem, and will be discussed separately.

Tape noise arises because of the essentially granular, or particulate, nature of the coating material and the magnetic domains within it. There are other effects, though, which are due to the actual recording process,

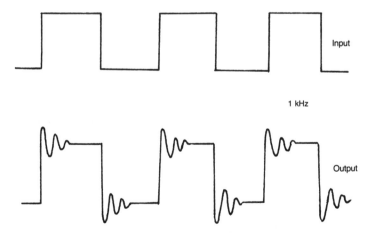

Fig. 1.17 *'Ringing' effects, on a 1 kHz square-wave, due to excessive HF boost applied to extend high-frequency response.*

and these are known as 'modulation' or 'bias noise', and 'contact' noise —
due to the surface characteristics of the tape.

The problem of noise on magnetic tape and that of the graininess of the
image in the photographic process are very similar. In both cases this is a
statistical problem, related to the distribution of the individual signal
elements, which becomes more severe as the area sampled, per unit time,
is decreased. (The strict comparison should be restricted to cine film, but
the analogy also holds for still photography.)

Similarly, a small output signal from the replay head, (equivalent to a
small area photographic negative), which demands high amplification,
(equivalent to a high degree of photographic enlargement), will lead to a
worse S/N ratio, (equivalent to worse graininess in the final image), and a
worse HF response and transient definition, (image sharpness).

For the medium itself, in photographic emulsions high sensitivity is
associated with worse graininess, and vice versa. Similarly, with magnetic
tapes, low noise, fine grain tape coatings also show lower sensitivity in
terms of output signal. Metal tapes offer the best solution here.

A further feature, well-known to photographers, is that prints from
large negatives have a subtlety of tone and gradation which is lacking
from apparently identical small-negative enlargements. Similarly, comparable size enlargements from slow, fine-grained, negative material are
preferable to those from higher speed, coarse-grained stock.

There is presumably an acoustic equivalence in the tape recording field.

Modulation or bias noise arises because of the random nature and
distribution of the magnetic material throughout the thickness of the
coating. During recording, the magnetic flux due to both the signal and
bias waveforms will redistribute these domains, so that, in the absence of
a signal, there will be some worsening of the noise figure. In the presence
of a recorded signal, the magnitude of the noise will be modulated by the
signal.

This can be thought of simply as a consequence of recording the signal
on an inhomogeneous medium, and gives rise to the description as 'noise
behind the signal', as when the signal disappears, this added noise also
stops.

The way in which the signal to noise ratio of a recorded tape is
dependent on the area sampled, per unit time, can be seen from a
comparison between a fast tape speed dual-track reel-to-reel recording
with that on a standard cassette.

For the reel-to-reel machine there will be a track width of 2.5 mm, with
a tape speed of 15 in./s (381 mm/s), which will give a sampled area
equivalent to 2.5 mm × 381 mm = 925 mm^2/s. In the case of the cassette
recorder the tape speed will be 1.875 in./s (47.625 mm/s), and the track
width will be 0.61 mm, so that the sampled area will only be 29 mm^2/s.
This is approximately 32 times smaller, equivalent to a 15 dB difference
in S/N ratio.

This leads to typical maximum S/N ratios of 67 dB for a two-track reel-to-reel machine, compared with a 52 dB value for the equivalent cassette recorder, using similar tape types. In both cases some noise reduction technique will be used, which will lead to a further improvement in these figures.

It is generally assumed that an S/N ratio of 70−75 dB for wideband noise is necessary for the noise level to be acceptable for high quality work, though some workers urge the search for target values of 90 dB or greater. However, bearing in mind that the sound background level in a very quiet domestic listening room is unlikely to be less than +30 dB (reference level, 0 dB = 0.0002 dynes/cm^2) and that the threshold of pain is only +105−110 dB, such extreme S/N values may not be warranted.

In cassette recorders, it is unusual for S/N ratios better than 60 dB to be obtained, even with optimally adjusted noise reduction circuitry in use.

The last source of noise, contact noise, is caused by a lack of surface smoothness of the tape, or fluctuations in the coating thickness due to surface irregularities in the backing material. It arises because the tape coating completes the magnetic circuit across the gap in the heads during the recording process.

Any variations in the proximity of tape to head will, therefore, lead to fluctuations in the magnitude of the replayed signal. This kind of defect is worse at higher frequencies and with narrower recording track widths, and can lead to drop-outs (complete loss of signal) in severe cases.

This kind of broadband noise is influenced by tape material, tape storage conditions, tape tension and head design. It is an avoidable nuisance to some extent, but is invariably present.

Contact noise, though worse with narrow tapes and low tape speeds, is not susceptible to the kind of analysis shown above, in relation to the other noise components.

The growing use of digital recording systems, even in domestic cassette recorder form, is likely to alter substantially the situation for all forms of recorded noise, although it is still argued by some workers in this field that the claimed advantages of digitally encoded and decoded recording systems are offset by other types of problem.

There is no doubt that the future of all large-scale commercial tape recording will be tied to the digital system, if only because of the almost universal adoption by record manufacturers of the compact disc as the future style of gramophone record.

This is recorded in a digitally encoded form, so it is obviously sensible to carry out the basic mastering in this form also, since their other large-scale products, the compact cassette and the vinyl disc, can equally well be mastered from digital as from analogue tapes.

ELECTRONIC CIRCUIT DESIGN

The type of circuit layout used for almost all analogue (i.e. non-digital) magnetic tape recorders is as shown, in block diagram form, in Fig. 1.18.

The critical parts of the design, in respect of added circuit noise, are the replay amplifier, and the bias/erase oscillator. They will be considered separately.

It should also be noted that the mechanical design of both the erase and record heads can influence the noise level on the tape. This is determined by contact profile and gap design, consideration of which is beyond the scope of this chapter.

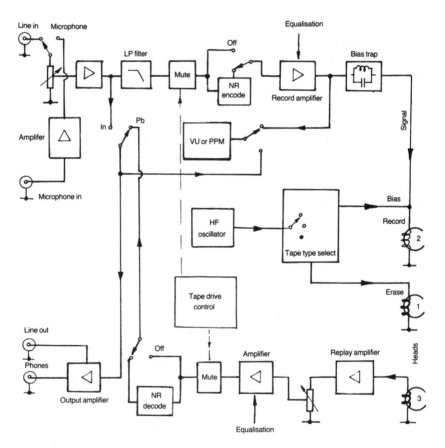

Fig. 1.18 *Schematic layout of the 'building blocks' of a magnetic tape recorder.*

The replay amplifier

Ignoring any noise reduction decoding circuitry, the function of this is to amplify the electrical signal from the replay head, and to apply any appropriate frequency response equalisation necessary to achieve a substantially uniform frequency response. The design should also ensure the maximum practicable s/n ratio for the signal voltage present at the replay head.

It has been noted previously that the use of very narrow replay head gaps, desirable for good high frequency and transient response, will reduce the signal output, if only because there will be less tape material within the replay gap at any one time. This makes severe demands on the design of the replay amplifier system, especially in the case of cassette recorder systems, where the slow tape speed and narrow tape track width reduces the extent of magnetic induction.

In practice, the mid-frequency output of a typical high quality cassette recorder head will only be of the order of 1 mV, and, if the overall S/N ratio of the system is not to be significantly degraded by replay amplifier noise, the 'noise floor' of the replay amp. should be of the order of 20 dB better than this. Taking 1 mV as the reference level, this would imply a target noise level of -72 dB if the existing tape S/N ratio is of the order of -52 dB quoted above.

The actual RMS noise component required to satisfy this criterion would be 0.25 μV, which is just within the range of possible circuit performance, provided that care is taken with the design and the circuit components.

The principal sources of noise in an amplifier employing semiconductor devices are:

1. 'Johnson' or 'thermal' noise, caused by the random motion of the electrons in the input circuit, and in the input impedance of the amplifying device employed (minimised by making the total input circuit impedance as low as possible, and by the optimum choice of input devices)

2. 'shot' noise, which is proportional to the current flow through the input device, and to the circuit bandwidth

3. 'excess', or '$1/f$', noise due to imperfections in the crystal lattice, and proportional to device current and root bandwidth, and inversely proportional to root frequency

4. collector-base leakage current noise, which is influenced by both operating temperature and DC supply line voltage

5. surface recombination noise in the base region.

Where these are approximately calculable, the equations shown below are appropriate.

$$\text{Johnson (thermal noise)} = \sqrt{4KTR\Delta f}$$

$$\text{Shot noise} = \sqrt{2qI_{DC} \times \Delta f}$$

$$\text{Modulation } (1/f) \text{ noise} = \frac{\sqrt{\Delta I \times \Delta f}}{f}$$

where 'Δf' is the bandwidth (Hz), $K = 1.38 \times 10^{-23}$, T is the temperature (Kelvin), q the electronic charge (1.59×10^{-19} Coulombs), f is the operating frequency and R the input circuit impedance.

In practical terms, a satisfactory result would be given by the use of silicon bipolar epitaxial-base junction transistor as the input device, which should be of PNP type. This would take advantage of the better surface recombination noise characteristics of the n-type base material, at an appropriately low collector to emitter voltage, say 3–4 V, with as low a collector current as is compatible with device performance and noise figure, and a base circuit impedance giving the best compromise between 'Johnson' noise and device noise figure requirements.

Other devices which can be used are junction field-effect transistors, and small-signal V-MOS and T-MOS insulated-gate type transistors. In some cases the circuit designer may adopt a slightly less favourable input configuration, in respect of S/N ratio, to improve on other performance characteristics such as slew-rate balance or transient response.

Generally, however, discrete component designs will be preferred and used in high quality machines, although low-noise integrated circuit devices are now attaining performance levels where the differences between IC and discrete component designs are becoming negligible.

Typical input amplifier circuit designs by Sony, Pioneer, Technics, and the author, are shown in Figs 1.19–1.22.

In higher speed reel-to-reel recorders, where the signal output voltage from the replay head is higher, the need for the minimum replay amplifier noise is less acute, and low-noise IC op. amplifiers, of the type designed for audio purposes will be adequate. Examples are the Texas Instruments TL071 series, the National Semiconductor LF351 series op amps. They offer the advantage of improved circuit symmetry and input headroom.

If the replay input amplifier provides a typical signal output level of 100 mV RMS, then all the other stages in the signal chain can be handled by IC op. amps without significant degradation of the S/N ratio. Designer preferences or the desire to utilise existing well tried circuit structures may encourage the retention of discrete component layouts, even in modern equipment.

However, in mass-produced equipment aimed at a large, low-cost,

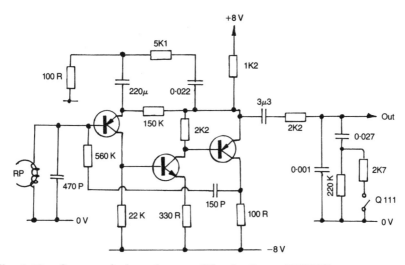

Fig. 1.19 *Cassette deck replay amplifier by Sony (TCK81).*

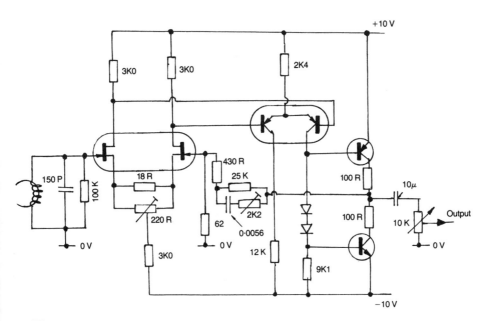

Fig. 1.20 *Very high quality cassette recorder replay amplifier due to Pioneer (CT-A9).*

market sale, there is an increasing tendency to incorporate as much as possible of the record, replay, and noise reduction circuitry within one or two complex, purpose-built ICs, to lessen assembly costs.

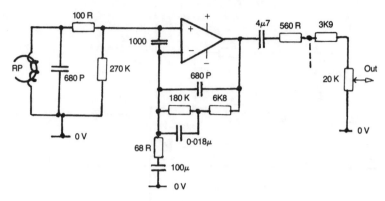

Fig. 1.21 *Combined record/replay amplifier used by Technics, shown in replay mode (RSX20).*

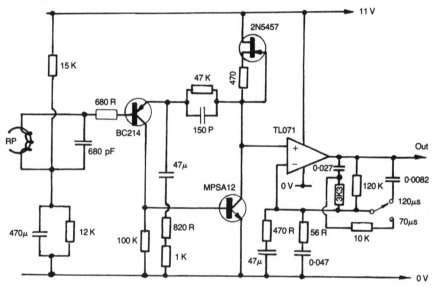

Fig. 1.22 *Complete cassette recorder replay system, used in portable machine.*

REPLAY EQUALISATION

This stage will usually follow the input amplifier stage, with a signal level replay gain control interposed between this and the input stage, where it will not degrade the overall noise figure, it will lessen the possibility of voltage overload and 'clipping' in the succeeding circuitry.

Opinions differ among designers on the position of the equalisation network. Some prefer to treat the input amplifier as a flat frequency response stage, as this means the least difficulty in obtaining a low input thermal noise figure. Others utilise this stage for the total or initial frequency response shaping function.

In all cases, the requirement is that the equalised replay amplifier response shall provide a uniform output frequency response, from a recorded flux level having the characteristics shown in Fig. 1.8, ignoring for the moment any frequency response modifications which may have been imposed by the record-stage noise reduction processing elements, or the converse process of the subsequent decoding stage.

Normally, the frequency response equalisation function will be brought about by frequency sensitive impedance networks in the feedback path of an amplifier element, and a typical layout is shown in Fig. 1.23.

Considering the replay curve shown in Fig. 1.24, this can be divided into three zones, the flattening of the curve, in zone 'A', below 50 Hz, in accordance with the LF 3180 µs time-constant employed, following Table 1.3, in almost all recommended replay characteristics, the levelling of the curve, in zone 'B', beyond a frequency determined by the HF time constant of Table 1.3, depending on application, tape type, and tape speed.

Since the output from the tape for a constant remanent flux will be assumed to increase at a +6 dB/octave rate, this levelling-off of the replay curve is equivalent to a +6 dB/octave HF boost. This is required to correct for the presumed HF roll-off on record, shown in Fig. 1.8.

Finally, there is the HF replay boost of zone 'C', necessary to compensate for poor replay head HF performance. On reel-to-reel recorders this effect may be ignored, and on the better designs of replay head in

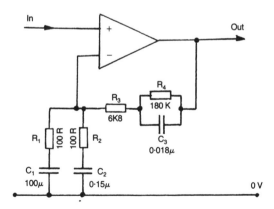

Fig. 1.23 *RC components network used in replay amplifier to achieve frequency response equalisation.*

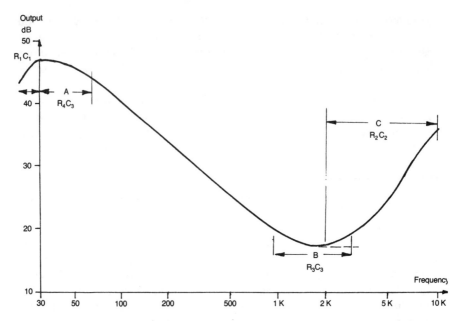

Fig. 1.24 *Cassette recorder replay response given by circuit layout shown in Fig. 1.23, indicating the regions affected by the RC component values.*

cassette recorders the necessary correction may be so small that it can be accomplished merely by adding a small capacitor across the replay head, as shown in the designs of Figs 1.19–1.22. The parallel resonance of the head winding inductance with the added capacitor then gives an element of HF boost.

With worse replay head designs, some more substantial boost may be needed, as shown in 'C' in Fig. 1.23.

A further type of compensation is sometimes employed at the low-frequency end of the response curve and usually in the range 30–80 Hz, to compensate for any LF ripple in the replay response, as shown in Fig. 1.10. Generally all that is attempted for this kind of frequency response flattening is to interpose a tuned LF 'notch' filter, of the kind shown in Fig. 1.25 in the output of the record amp, to coincide with the first peak in the LF replay ripple curve.

BIAS OSCILLATOR CIRCUITS

The first major requirement for this circuit is a substantial output voltage swing – since this circuit will also be used to power the 'erase' head, and with some tapes a large erase current is necessary to obtain a low residual

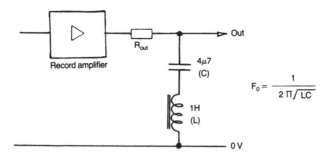

$$F_0 = \frac{1}{2\,\Pi\sqrt{LC}}$$

Fig. 1.25 *Notch filter in record amplifier, used to lessen LF ripple effects.*

signal level on the erased tape. Also needed are good symmetry of waveform, and a very low intrinsic noise figure, since any noise components on the bias waveform will be added to the recorded signal.

The invariable practical choice, as the best way of meeting these requirements, is that of a low distortion HF sine waveform.

Since symmetry and high output voltage swing are both most easily obtained from symmetrical 'push–pull' circuit designs, most of the better modern recorders employ this type of layout, of which two typical examples are shown in Figs 1.26 and 1.27. In the first of these, which is used in a simple low-cost design, the erase coil is used as the oscillator inductor, and the circuit is arranged to develop a large voltage swing from a relatively low DC supply line.

The second is a much more sophisticated and ambitious recorder design. Where the actual bias voltage level is automatically adjusted by micro-computer control within the machine to optimise the tape recording characteristics. The more conventional approach of a multiple winding transformer is employed to generate the necessary positive feedback signal to the oscillator transistors.

In all cases great care will be taken in the choise of circuitry and by the use of high 'Q' tuned transformers to avoid degrading the purity of the waveform or the S/N ratio.

In good quality commercial machines there will generally be both a switch selectable bias level control, to suit the IEC designated tape types (1–4) and a fine adjustment. This latter control will usually be a pair of variable resistors, in series with separate small capacitors, in the feed to each recording channel, since the high AC impedance of the capacitor tends to restrict the introduction of lower frequency noise components into the record head. It also serves to lessen the possibility of recording signal cross-talk from one channel to the other.

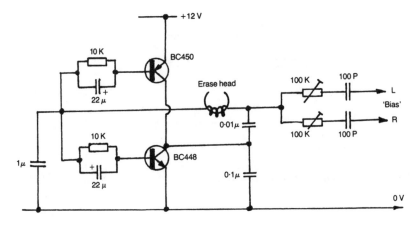

Fig. 1.26 *High-output, low-distortion, bias/erase oscillator used in cassette recorder.*

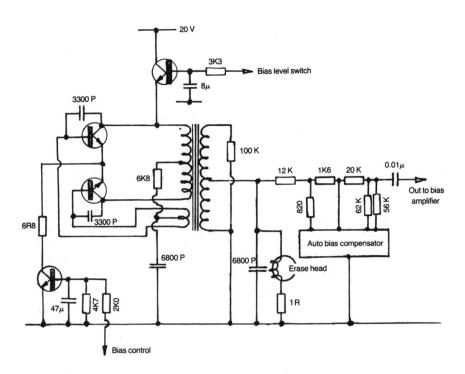

Fig. 1.27 *High-quality bias oscillator having automatic bias current adjustment facilities (Pioneer CT-A9).*

THE RECORD AMPLIFIER

Referring to the block diagram of Fig. 1.18, the layout of the recording amplifier chain will normally include some noise reduction encoding system (Dolby A or equivalent, and/or Dolby HX, in professional analogue-type tape recorders, or Dolby B or C in domestic machines). The operation of this would be embarrassed by the presence of high-frequency audio components, such as the 19 kHz FM 'pilot tone' used in stereo broadcasts to regenerate the 38 kHz stereo sub-carrier.

For this reason, machines equipped with noise reduction facilities will almost invariably also include some low-pass filter system, either as some electronic 'active filter' or a simple LC system, of the forms shown in Fig. 1.28. Some form of microphone amplifier will also be provided as an alternative to the line inputs. The design of this will follow similar low-noise principles to those essential in the replay head amplifier, except that the mic. amplifier will normally be optimised for a high input circuit impedance. A typical example is shown in Fig. 1.29.

As mentioned above, it is common (and good) practice to include a 'muting' stage in the replay amp line, to restrict the audibility of the amplifier noise in the absence of any tape signal. A similar stage may be incorporated in the record line, but in this case the VU or peak recording meter will require to derive its signal feed from an earlier stage of the amplifier. Thus the input signal level can still be set with the recorder at 'pause'.

The recording amplifier will be designed to deliver the signal to the record head at an adequate amplitude, and with low added noise or distortion, and in a manner tolerant of the substantial HF bias voltage present at the record head.

Contemporary design trends tend to favour record heads with low winding impedances, so the record amplifier must be capable of coping with this.

As noted above, the design of the record amplifier must be such that it will not be affected by the presence of the HF bias signal at the record head.

Several approaches can be chosen. One is to employ an amplifier with a very high output impedance, the output signal characteristics of which will not be influenced by superimposed output voltage swings. Another is to interpose a suitable high value resistor in the signal line between the record amp. and the head. This also has the benefit of linearising the record current versus frequency characteristics of the record system. If the record amplifier has a low output impedance, it will cause the intruding bias frequency signal component to be greatly attenuated in its reverse path.

However, the simplest, and by far the most common, choice is merely

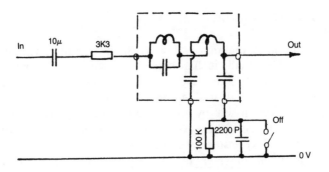

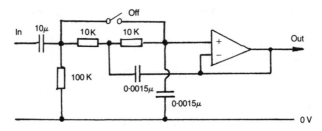

Fig. 1.28   *LC and active low-pass filter systems used to protect noise reduction circuitry from spurious HF inputs.*

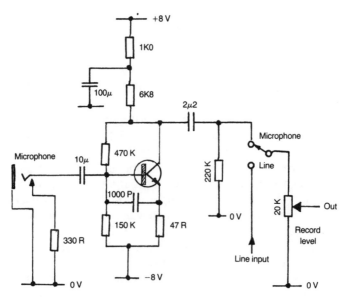

Fig. 1.29   *Single stage microphone input amplifier by Sony (TCK-81).*

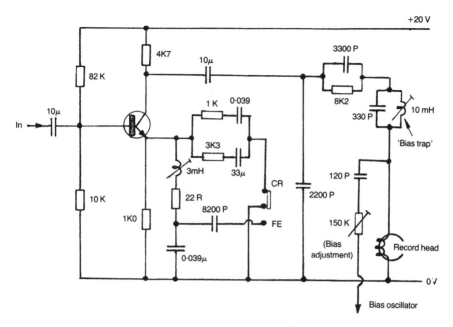

Fig. 1.30 *Record output stage, from Pioneer, showing bias trap and LC equalisation components (CT4141-A).*

to include a parallel-tuned LC rejection circuit, tuned to the HF bias oscillator frequency, in the record amplifier output, as shown in the circuit design of Fig. 1.30.

Some very advanced recorder designs actually use a separate mixer stage, in which the signal and bias components can be added before they are fed to the record head. In this case the design of the output stage is chosen so that it can handle the total composite (signal + bias) voltage swing without cross-modulation or other distortion effects.

RECORDING LEVEL INDICATION

This is now almost exclusively performed in domestic type equipment by instantaneously acting light-emitting diodes, adjusted to respond to the peak recording levels attained. There are several purpose-built integrated circuits designed especially for this function.

In earlier instruments, where moving coil display meters were used, great care was necessary to ensure that the meter response had adequate ballistic characteristics to respond to fast transient signal level changes. This invariably meant that good performance was expensive.

It is notable that professional recording engineers still prefer to use meter displays, the ballistic characteristics of which are normally precisely controlled, and to the ASA C16-5 specification. However, the indication given by these may be supplemented by LED-type peak level indication.

TAPE DRIVE MOTOR SPEED CONTROL

Few things have a more conspicuous effect upon the final performance of a tape recorder system than variations in the tape speed. Electronic servo systems, of the type shown in schematic form in Fig. 1.31, are increasingly used in better class domestic machines (they have been an obligatory feature of professional recorder systems for many years) to assure constancy of tape travel.

In arrangements of this kind, the capstan drive motors will either be direct current types, the drive current of which is increased or decreased electronically depending on whether the speed sensor system detects a slowing or an increase in tape speed. Alternatively, the motors may be of 'brushless' AC type, fed from an electronically generated AC waveform, the frequency of which is increased or decreased depending on whether a positive or a negative tape speed correction is required.

With the growing use of microprocessor control and automated logic sequences to operate head position, signal channel and record/replay selection, and motor control, the use of electronically generated and regulated motor power supplies is a sensible approach. The only problem is that the increasing complexity of circuitry involved in the search for better performance and greater user convenience may make any repairs more difficult and costly to perform.

PROFESSIONAL RECORDING EQUIPMENT

This is aimed at an entirely different type of use to that of the domestic equipment, and is consequently of different design, with a different set of priorities.

Domestic tape cassette recorders are used most commonly for transferring programme material from radio broadcasts gramophone or other tape recordings, or spontaneous (i.e., non-repeatable) 'live' events, to tape and the finished recording is made for the benefit of the recordist or the participants. No editing is desired or practicable. In contrast, in commercial/professional work it is likely that the final product will be closely scrutinised for imperfections, and will be highly edited.

Moreover, in broadcasting or news applications, it is essential that time synchronisation, for example with pictures, be practicable.

Also, since the recording process is just one step in a necessarily

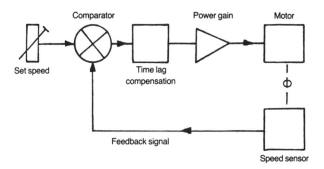

Fig. 1.31 *Schematic layout of speed control servo-mechanism.*

profitable commercial operation, professional machines must be rugged, resistant to misuse, and reliable. The cost of the equipment, except as dictated by competitive pressures between rival manufacturers, will be of a secondary consideration, and its physical bulk will be less important than its ease of use.

These differences in use and outlook are reflected in the different types of specification of professional recording systems, and these are listed below.

GENERAL DESCRIPTION

Mechanical

Professional tape recording machines will invariably be reel-to-reel recording as 'half-track', or occasionally as 'quarter-track' on ¼ in. tape, or as 8-track, 16-track, 24-track or 32-track, on ½ in., 1 in. or 2 in. wide tape. (Because of the strong influence of the USA in the recording field, metric dimensions are very seldom used in specifying either tape widths or linear tape speeds of professional equipment.)

Even on half-track machines, facilities are now being provided for time code marking as are more generally found on multi-track machines by way of an additional channel in the central 2.25 mm dead band between the two stereo tracks.

This allows identification of the time elapsed since the start of the recording as well as the time remaining on the particular tape in use. This information is available even if the tape is removed and replaced, part wound. (It is customary for professional recorders, copied now by some of the better domestic machines, to indicate tape usage by elapsed-time displays rather than as a simple number counter.) Also available is the real time − as in the conventional clock sense − and any reference markers needed for external synchronisation or tape editing.

The mechanisms used will be designed to give an extremely stable tape tension and linear tape speed, with very little wow or flutter, a switched choice of speeds, and provision for adjusting the actual speed, upwards or downwards, over a range − customarily chosen as seven semitones − with facility for accurate selection of the initial reference tape speed. It is also customary to allow for different sized tape spools, with easy and rapid replacement of these.

The high power motors customarily used to ensure both tape speed stability and rapid acceleration − from rest to operating speed in 100 ms is attainable on some machines − can involve significant heat dissipations. So the bed-plate for the deck itself is frequently a substantial die-casting, of up to two inches in thickness, in which the recesses for the reels, capstans and heads are formed by surface machining . The use of ancillary cooling fans in a studio or sound recording environment is understandably avoided wherever possible.

To make the tape speed absolutely independent of mains power frequency changes, it is common practice to derive the motor speed reference from a crystal controlled source, frequently that used in any microprocessor system used for other control, editing, and search applications.

Since the product will often be a commercial gramophone record or cassette recording, from which all imperfections must be removed before production, editing facilities − either by physically splicing the tape, or by electronic processing − will feature largely in any 'pro' machine. Often the tape and spool drive system will permit automatic 'shuttle' action, in which the tape is moved to and fro across the replay head, to allow the precise location of any edit point.

An internal computer memory store may also be provided to log the points at which edits are to be made, where these are noted during the recording session. The number and position of these may also be shown on an internal display panel, and the tape taken rapidly to these points by some auto-search facility.

Electrical

The electronic circuitry employed in both professional and domestic tape recorder systems is similar, except that their tasks are different. Indeed many of the facilities which were originally introduced in professional machines have been taken over into the domestic scene, as the ambitions of the manufacturers and the expectations of the users have grown.

In general outline, therefore, the type of circuit layout described above will be appropriate. However, because of the optional choice of tape speeds, provision will be made for different, switch-selectable, equalisation characteristics and also, in some cases, for different LF compensation to offset the less effective LF recording action at higher tape speeds. Again, the bias settings will be both adjustable and pre-settable to suit the tape types used.

To assist in the optimisation of the bias settings, built-in reference oscillators, at 1 kHz, 10 kHz, and sometimes 100 Hz, may be provided, so that the flattest practicable frequency response may be obtained.

This process has been automated in some of the more exotic cassette recorders, such as the Pioneer CT-A9, in which the bias and signal level is automatically adjusted when the recording is initiated, using the first eleven seconds worth of tape, after which the recorder resets itself to the start of the tape to begin recording. Predictably, this style of thing does not appeal greatly to the professional recording engineer!

Whether or not noise reduction circuitry will be incorporated within the machine is a matter of choice. 'Pro' equipment will, however, offer easy access and interconnection for a variety of peripheral accessories, such as remote control facilities and limiter and compressor circuitry, bearing in mind the extreme dynamic range which can be encountered in live recording sessions.

Commonly, also, the high-frequency headroom expansion system, known as the Dolby HX Pro will be incorporated in high quality analogue machines. This monitors the total HF components of the recorded signal, including both programme material and HF bias, and adjusts the bias voltage output so that the total voltage swing remains constant. This can allow up to 6 dB more HF signal to be recorded without tape saturation.

In the interests of signal quality, especially at the high frequency end of the audio spectrum, the bias frequency employed will tend to be higher than in lower tape speed domestic machines. 240 kHz is a typical figure.

However, to avoid overheating in the erase head, caused by the eddy-current losses which would be associated with such a high erase frequency, a lower frequency signal is used, such as 80 kHz, of which the bias frequency employed is the third harmonic. This avoids possible beat frequency effects.

Very frequently, both these signals are synthesised by subdivision from the same crystal controlled master oscillator used to operate the internal microprocessor and to control the tape drive speed.

With regard to the transient response shortcomings of tape recording systems, referred to earlier, there is a growing tendency in professional analogue machines to employ HF phase compensation, to improve the step-function response. The effect of this is illustrated in Fig. 1.32, and the type of circuitry used is shown in Fig. 1.33.

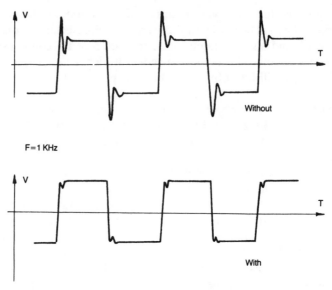

F=1 KHz

Fig. 1.32 *Improvement in HF transient response, and consequent stereo image quality, practicable by the use of HF phase compensation.*

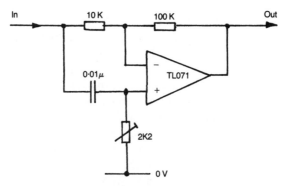

Fig. 1.33 *HF phase-compensation circuit.*

A major concern in professional work is the accurate control of both recording and output levels. Frequently the recording level meters will have switch selectable sensitivities for their OVU position, either as the basic standard (OVU = 1 mW into 600 ohms = 0.774 V RMS) or as 'dB V' (referred to 1 V RMS). It is also expected that the OVU setting will be related accurately to the magnetisation of the tape, with settings selectable at either 185, 250 or 320 nanoWebers/m.

Finally, and as a clear distinction from amateur equipment, professional units will all have provision for balanced line inputs from microphone and other programme sources, since this greatly lessens the susceptibility of such inputs to mains induced hum and interference pick-up.

MULTI-TRACK MACHINES

Although single channel stereo machines have a place in the professional recording studio, the bulk of the work will be done on multi-track machines, either 8-, 16-, 24-, or 32-track. The reasons for this are various.

In the pop field, for example, it is customary for the individual instrumentalists or singers in a group to perform separately in individual soundproofed rooms, listening to the total output, following an initial stereo mixing, on 'cans' (headphones). Each individual contributor to the whole is then recorded on a separate track, from which the final stereo master tape can be recorded by mixing down and blending, with such electronic sound effects as are desired.

This has several advantages, both from the point of view of the group, and for the recording company – who may offer a sample of the final product as a cassette, while retaining the master multi-track tape until all bills are paid. For the performers, the advantages are that the whole session is not spoilt by an off-colour performance by any individual within the group, since this can easily be re-recorded. Also, if the number is successful, another equally successful 'single' may perhaps be made, for example, by a re-recording with a different vocal and percussion section.

Moreover, quite a bit of electronic enhancement of the final sound can be carried out after the recording session, by frequency response adjustment and added reverberation, echo and phase, so that fullness and richness can be given, for instance, to an otherwise rather thin solo voice. Studios possessing engineers skilled in the arts of electronic 'fudging' soon become known and sought after by the aspiring pop groups, whose sound may be improved thereby.

On a more serious level, additional tracks can be invaluable for recording cueing and editing information. They are also useful in the case, say, of a recording of a live event, such as an outside broadcast, or a symphony orchestra, in gathering as many possible signal inputs, from distributed microphones, as circumstances allow. The balance between these can then be chosen, during and after the event, to suit the intended use. A mono radio broadcast needing a different selection and mix of signal than, for example, a subsequent LP record, having greater dynamic range.

This 'multi-miking' of large scale music is frowned upon by some recording engineers. Even when done well, it may falsify the tonal scale of solo instrumentalists, and when done badly may cause any soloist to have an acoustic shadow who follows his every note, but at a brief time interval later. It has been claimed that the 'Sound Field' system will do all that is necessary with just a simple tetrahedral arrangement of four cardioid microphones and a four-channel recorder.

A technical point which particularly relates to multi-track recorder systems is that the cross-talk between adjacent channels should be very low. This will demand care in the whole record/replay circuitry, not just the record/replay head. Secondly, the record head should be usable, with adequate quality, as a replay head.

This is necessary to allow a single instrumentalist to record his contribution as an addition to a tape containing other music, and for it to be recorded in synchronism with the music which he hears on the monitor headphones, without the record head to replay head tape travel time lag which would otherwise arise.

The need to be able to record high-frequency components in good phase synchronism, from channel to channel, to avoid any cancellation on subsequent mixing, makes high demands on the vertical alignment of the record and replay head gaps on a multi-channel machine. It also imposes constraints on the extent of wander or slewing of the tape path over the head.

DIGITAL RECORDING SYSTEMS

The basic technique of digitally encoding an analogue signal entails repetitively sampling the input signal at sufficiently brief intervals, and then representing the instantaneous peak signal amplitude at each successive moment of sampling as a binary coded number sequence consisting of a train of '0s' and '1s'. This process is commonly termed 'pulse code modulation', or 'PCM'.

The ability of the encoding process to resolve fine detail in the analogue waveform ultimately depends on the number of 'bits' available to define the amplitude of the signal, which, in turn, determines the size of the individual step in the 'staircase' waveform, shown in Fig. 1.34, which results from the eventual decoding of the digital signal.

Experience in the use of digitally encoded audio signals suggests that the listener will generally be unaware of the digital encoding/decoding process if '16-bit' (65 536 step) resolution is employed, and that resolution levels down to '13-bit' (8192 steps), can give an acceptable sound quality in circumstances where some low-level background noise is present, as, for example, in FM broadcast signals.

The HF bandwidth which is possible with a digitally encoded signal is determined by the sampling frequency, and since the decoding process is unable to distinguish between signals at frequencies which are equally

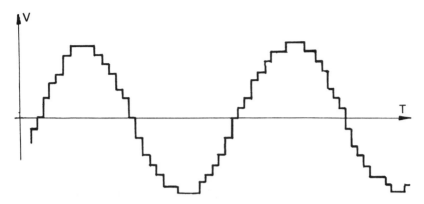

Fig. 1.34 *'Staircase' type waveform resulting from the digital encoding/ decoding process. (1300 Hz sine-waveform illustrated as sampled at 44.1 kHz, and at 4-bit resolution).*

spaced above and below half the sampling frequency, steep-cut low-pass filters must be used both before and after the digital encoding/decoding stages if spurious audio signals are to be avoided.

In the case of the compact disc, a sampling frequency of 44.1 kHz is used. This allows a HF turn-over frequency of 20 kHz, by the time 'anti-aliasing' filtering has been included. In the case of the '13-bit' pulse-code modulated distribution system used by the BBC for FM stereo broadcasting, where a sampling frequency of 32 kHz is used, the 'aliasing' frequency is 16 kHz, which limits the available HF bandwidth to 14.5 kHz.

It must be noted that the digital encoding (PCM) process suffers from the disadvantage of being very extravagant in its use of transmission bandwidth. For example, in the case of the 'compact disc', by the time error correction 'parity bits' and control signals have been added, the digital pulse train contains 4 321 800 bit/s. Even the more restricted BBC PCM system requires a minimum channel bandwidth of 448 kHz. Obviously, the use of a wider audio pass-band, or a greater resolution, will demand yet higher bit rates.

If it is practicable to provide the wide HF bandwidth required, the use of digitally encoded signals offers a number of compelling advantages. Of these probably the most important is the ability to make an unlimited number of copies, each identical to the 'first generation' master tape, without any copy to copy degradation, and without loss of high frequency and low-level detail from the master tape due to flux induced demagnetis-ation during the replay process.

In addition, the digital process allows a ruler flat replay response from, typically, 5 Hz–20 kHz, and offers signal to noise ratio, dynamic range, channel separation, frequency stability and harmonic and intermodulation distortion characteristics which are greatly superior to those offered by even the best of the traditional analogue reel-to-reel equipment.

Moreover, digital '0/1' type encoding shows promise of greater archival stability, and eliminates 'print-through' during tape storage, while the use of 'parity bit' (properly known as 'cross interleave Reed-Solomon code') error correction systems allow the correction or masking of the bulk of the faults introduced during the recording process, due, for example, to tape 'drop outs' or coating damage.

Specific bonuses inherent in the fact that the PCM signal is a data stream (not in the time domain until divided into time-related segments at intervals determined by some master frequency standard) are the virtual absence of 'wow', 'flutter' and tape speed induced errors in pitch, and also that the signals can be processed to remove any time or phase errors between tracks.

To illustrate the advantages of digitally encoded signal storage, a comparison between the performance of a good quality analogue recording, for example, a new (i.e., not previously played) 12 inch vinyl 'LP' disk, and a 16-bit digitally encoded equivalent is shown in Table 1.5.

There are, of course, some snags. Of these, the most immediate is the problem of 'editing' the digitally encoded signal on the tape. In the case of conventional analogue reel-to-reel tape machines, editing the tape to remove faults, or to replace unsatisfactory parts of the recording, is normally done by the recording engineer by inching the tape, manually,

Table 1.5 Analogue versus digital performance characteristics

	12″ LP	16-bit PCM system (i.e., compact disc)
Frequency response	30 Hz−20 kHz +/−2 dB	5 Hz−20 kHz +/−0.2 dB
Channel separation	25−30 dB	>100 dB
S/N ratio	60−65 dB	>100 dB
Dynamic range	Typically 55 dB @1 kHz	>95 dB
Harmonic distortion	0.3−1.5%	<0.002%
'Wow' and 'flutter'	0.05%	nil
Bandwidth requirement	20 kHz	4.3 MHz

backwards and forwards across the general region of the necessary edit, while listening to the output signal, until the precise cut-in point is found. A physical tape replacement is then made by 'cut and splice' techniques, using razor blade and adhesive tape.

Obviously, editing a digital tape is, by its nature, a much more difficult process than in the case of an analogue record, and this task demands some electronic assistance. A range of computer-aided editing systems has therefore been evolved for this purpose, of which the preferred technique, at least for '16-bit' 'stereo' recordings, is simply to transfer the whole signal on to a high capacity (1 Gbyte or larger) computer 'hard disk', when the edit can be done within the computer using an appropriate programme.

The second difficulty is that the reproduced waveform is in the form of a 'staircase' (as shown in the simplified example shown in Fig. 1.34), and this is a characteristic which is inherent in the PCM process. The type of audible distortion which this introduces becomes worse as the signal level decreases, and is of a curious character, without a specific harmonic relationship to the fundamental waveform.

The PCM process also causes a 'granularity' in low level signals, most evident as a background noise, called 'quantisation noise', due to the random allocation of bit level in signals where the instantaneous value falls equally between two possible digital steps. This kind of noise is always present with the signal and disappears when the signal stops: it differs in this way from the noise in an analogue recording, which is present all the time.

In normal PCM practice, a random 'dither' voltage, equal to half the magnitude of the step, is added to the recorded signal, and this increases the precision in voltage level resolution which is possible with signals whose duration is long enough to allow many successive samples to be taken at the same voltage level. When the output signal is then averaged, by post-decoder filtering, much greater subtlety of resolution is possible, and this degree of resolution is increased as the sampling frequency is raised.

This increase in the possible subtlety of resolution has led to the growing popularity of 'oversampling' techniques in the replay systems of 'CD' players (a process which is carried to its logical conclusion in the 256× oversampling 'bitstream' technique, equivalent to a 11.29 MHz sampling rate), although part of the acoustic benefit given by increasing the sampling rate is simply due to the removal of the need for the very steep-cut, 20 kHz, anti-aliasing output filters essential at the original 44.1 kHz sampling frequency. Such so-called 'brick wall' filters impair transient performance, and can lead to a 'hard' or 'over bright' treble quality.

Some critical listeners find, or claim to find, that the acoustic characteristics of 'digital' sound are objectionable. They are certainly clearly detectable at low encoding resolution levels, and remain so, though less conspicuously, as the resolution level is increased.

For example, if the listener is allowed to listen to a signal encoded at 8-bit resolution, and the digital resolution is then progressively increased, by stages, to 16-bits, the 'digital' character of the sound still remains noticeable, once the ear has been alerted to the particular nature of the defects in reproduced signal. This fault is made much less evident by modern improvements in replay systems.

Professional equipment

So far as the commercial recording industry is concerned, the choice between analogue and digital tape recording has already been made, and recording systems based on digitally encoded signals have effectively superseded all their analogue equivalents.

The only exceptions to this trend are in small studios, where there may seem little justification for the relatively high cost of replacing existing analogue recording equipment, where this is of good quality and is still fully serviceable. This is particularly true for 'pop' music, which will, in any case, be 'mixed down' and extensively manipulated both before and after recording, by relatively low precision analogue circuitry.

Some contemporary professional digital tape recording systems can offer waveform sampling resolution as high as '20-bit' (1 048 576 possible signal steps), as compared with the '16-bit' encoding (65 536 steps) used in the compact disc system. The potential advantages, at least for archive purposes, offered by using a larger number of steps in the digital 'staircase' are that it reduces the so-called 'granularity' of the reproduced sound, and it also reduces the amount of 'quantisation noise' associated with the digital–analogue decoding process.

Unfortunately, in the present state of the art, part of the advantage of such high resolution recording will be lost to the listener since the bulk of such recordings will need to be capable of a transfer to compact disc, and, for this transfer, the resolution must be reduced once again to '16-bit', with the appropriate resolution enhancing sub-bit 'dither' added once more to the signal.

Some 'top of the range' professional fixed head multi-track reel-to-reel digital recorders can also provide on-tape cueing for location of edit points noted during recording, as well as tape identification, event timing and information on tape usage, shown on a built-in display system.

However, although multi-track fixed head recorders are used in some of the larger commercial studios, a substantial number of professional recording engineers still prefer the Sony 'U-matic' or '1630' processor, which is, effectively, a direct descendant of the original Sony 'PCM-1' and 'PCM-F1' audio transfer systems, which were designed to permit '16-bit' digitally encoded audio signals to be recorded and reproduced by way of a standard rotary head video cassette recorder.

It is unlikely that the situation in relation to digital recording will change significantly in the near future, except that the recording equipment will become somewhat less expensive and easier to use, with the possibility of semi-automatic editing based on 'real time' information recorded on the tape.

What does seem very likely is that variations of the 'near-instantaneous companding' (dynamic range compression and expansion) systems developed for domestic equipment will be employed to increase the economy in the use of tape, and enhance the apparent digital resolution of the less expensive professional machines.

Domestic equipment

In an ideal world, the only constraints on the design and performance of audio equipment intended for domestic use would be the need to avoid too high a degree of complexity − the cost of which might limit its saleability − or the limitations due simply to the physical characteristics of the medium.

However, in reality, although the detailed design of equipment offered by a mass-market manufacturer will be influenced by his technical competence, the basic function of the equipment will be the outcome of deliberations based on the sales forecasts made by the marketing divisions of the companies concerned, and of the policy agreements made behind the closed doors of the trade confederations involved.

A good example of the way in which these forces operate is given by the present state of 'domestic' (i.e., non-professional) cassette recording technology. At the time of writing (1994), it is apparent that the future for high-quality domestic tape recording systems lies, just as certainly as in the case of professional systems, in the use of some form of digitally encoded signal.

This is a situation which has been known, and for which the technical expertise has been available for at least 20 years, but progress has been virtually brought to a standstill by a combination of marketing uncertainties and objections from the record manufacturing companies.

As noted above, the major advantage offered by digital recording/ replay systems is that, in principle, an unlimited number of sequential copies can be made, without loss of quality − a situation which is not feasible with the 'analogue' process.

The realisation that it is now possible, by direct digital transfer of signals from a compact disc, to make a multiplicity of blemish free copies, indistinguishable from the original, has frightened the recording companies, who fear, quite reasonably, that this could reduce their existing large profits.

As a result, their trade organisation, the International Federation of Phonograph Industries (IFPI) has pressed for some means of preventing this, at least so far as the domestic user is concerned, and has brought pressure to bear on their national governments to prevent importation of digital cassette recorders until the possibility of unlimited digital copying has been prevented.

This pressure effectively put a stop to the commercial development, ten years ago, of digital audio tape (DAT) machines, apart from such pioneering units as the Sony PCM-F1, which is designed for use with standard ¾ inch rotary-head video tape recorders, and which were classified as 'professional' units.

A technical solution to the record manufacturers objections has now been developed, in the form of the 'Serial Copy Management System' (SCMS). On recorders fitted with this device, a digitally coded signal is automatically added to the 'first generation' tape copy, when this is made from a 'protected' source, such as a compact disc, and this will prevent any further, 'second generation', digital copies being made on an 'SCMS' equipped recorder. However, even if the copy is made from an unprotected source, such as a direct microphone input, a digital copy of this material will still carry a 'copy code' signal to prevent more than one further generation of sequential copies from being made.

In the interim period, the debate about the relative merits of domestic 'S-DAT' recorders (digital tape recorders using stationary head systems) and 'R-DAT' systems (those machines based on a rotary head system, of the kind used in video cassette recorders) had faded away. In its original form, 'S-DAT' is used solely on multiple track professional systems, and 'R-DAT' apparatus is offered, by companies such as Sony, Aiwa and Casio, mainly as portable recorder systems, which come complete with copy-code protection circuitry.

The success of any mass-market recording medium is dependent on the availability of pre-recorded music or other entertainment material, and because of the uncertainties during the pre-'SCMS' period, virtually no pre-recorded digital tapes (DAT) have been available, and now that the political problems associated with 'DAT' seem to have been resolved, three further, mutually competitive, non-professional recording systems have entered the field.

These are the recordable compact disc (CDR), which can be replayed on any standard CD player: the Sony 'Mini-Disc', which is smaller (3 in), and requires a special player: and the Philips 'Digital Compact Cassette' (DCC), which is a digital cassette recorder which is compatible with existing compact cassettes for replay purposes.

This latter equipment was developed as a result of a commercial prediction, by the Philips marketing division, that vinyl discs, especially 12 inch 'LP's, would no longer be manufactured beyond 1995, and that compact cassettes, at present the major popular recorded music medium, would reach a sales peak in the early 1990s, and would then begin to decline in sales volume, as users increasingly demanded 'CD' style replay quality.

Compact discs would continue their slow growth to the mid-1990s but sales of these would become static, from then onwards, mainly due to market saturation. What was judged to be needed was some simple and inexpensive apparatus which would play existing compact cassettes, but which would also allow digital recording and replay at a quality standard comparable with that of the compact disc.

The Philips answer to this perceived need is the digital compact cassette player, a machine which would accept, and replay, any standard compact cassette, and would therefore be 'backwards compatible' both with existing cassette stocks, and user libraries. However, when used as a recorder, the unit will both record and replay a digitally encoded signal, at the existing 4.76 cm/s tape speed, using a fixed eight-track head.

Unfortunately, digital recording systems are extravagant in their use of storage space, and the technical problems involved in trying to match the performance of a compact disc, and in then accommodating the digitally encoded signal on such a relatively low speed, narrow width tape, as that used in the compact cassette, are daunting. Even with the effective resolution reduced, on average, to the 4-bit level, a substantial reduction in bit rate is necessary from the 4.3218 Mbit/s of the compact disk to the 384 kbit/s possible with the digital compact cassette.

The technical solutions offered by Philips to this need for an elevenfold bit-rate reduction are what is termed 'precision adaptive sub-band coding' (PASC), and 'adaptive bit allocation' (ABA). The 'PASC' system is based on the division of the incoming audio signal into 32 separate frequency 'sub-bands', and then comparing the magnitude of the signal in each of these channels with the total instantaneous peak signal level. The sub-band signals are then weighted according to their position in the frequency spectrum, and any which would not be audible are discarded.

The 'ABA' process takes into account the masking effect on any signal of the presence of louder signals at adjacent, especially somewhat lower, frequencies, and then discards any signals, even when they are above the theoretical audibility threshold, if they are judged to be likely to be masked by adjacent ones. The 'spare' bits produced by this 'gardening' exercise are then re-allocated to increase the resolution available for other parts of the signal.

Reports on demonstrations of the digital compact cassette system which have been given to representatives of the technical press, suggest that Philips' target of sound quality comparable with that of the compact disc has been achieved, and the promised commercial backing from the record companies leads to the expectation that pre-recorded cassettes will be available, at a price intermediate between that of the CD and the existing lower quality compact cassette.

Blank tapes, using a chromium dioxide coating, will be available, initially, in the 'C60' and 'C90' recording lengths, and are similar in size to existing analogue cassettes, but have an improved mechanical construction, to allow a greater degree of precision in tape positioning.

In the analogue field, improved cassette record/replay heads capable of a genuine 20 Hz—20 kHz frequency response are now available, though tape characteristics, even using metal tape, prevent this bandwidth being attained at greater signal levels than 10 dB below the maximum output level (see pages 7–8, 15–25).

There is a greater availability, in higher quality cassette recorders, of the 'HX-PRO' bias control system, in which the HF signal components present in the incoming programme material are automatically taken into account in determining the overall bias level setting. When this is used in association with the Dolby 'C' noise reduction process s/n ratios better than 70 dB have been quoted.

A further improvement in performance on pre-recorded compact cassettes has been offered by Nakamichi by the use of their patented split-core replay head system, which is adjusted automatically in use to achieve the optimum azimuth alignment, for replaying tapes mastered on other recorders, and 'three-head' recording facilities are now standard in 'hi-fi' machines.

The extent to which manufacturers will continue to improve this medium is, however, critically dependent on the success, in the practical terms of performance, cost and reliability, of the digital compact cassette system. Whatever else, the future for domestic equipment is not likely to be one of technical stagnation.

CHAPTER 2

Tuners and radio receivers

BACKGROUND

The nineteenth century was a time of great technical interest and experiment in both Europe and the USA. The newly evolved disciplines of scientific analysis were being applied to the discoveries which surrounded the experimenters and which, in turn, led to further discoveries.

One such happening occurred in 1864, when James Clerk Maxwell, an Edinburgh mathematician, sought to express Michael Faraday's earlier law of magnetic induction in proper mathematical form. It became apparent from Maxwell's equations that an alternating electromagnetic field would give rise to the radiation of electromagnetic energy.

This possibility was put to the test in 1888 by Heinrich Hertz, a German physicist. He established that this did indeed happen, and radio transmission became a fact. By 1901, Marconi, using primitive spark oscillator equipment, had transmitted a radio signal in Morse code across the Atlantic. By 1922 the first commercial public broadcasts had begun for news and entertainment.

By this time, de Forrest's introduction of a control grid into Fleming's thermionic diode had made the design of high power radio transmitters a sensible engineering proposition. It had also made possible the design of sensitive and robust receivers. However, the problems of the system remain the same, and the improvements in contemporary equipment are merely the result of better solutions to these, in terms of components or circuit design.

BASIC REQUIREMENTS

These are

- **selectivity** − to be able to select a preferred signal from a jumble of competing programmes;
- **sensitivity** − to be sure of being able to receive it reliably;
- **stability** − to be able to retain the chosen signal during the required reception period;
- **predictability** − to be able to identify and locate the required reception channel;

- **clarity** – which requires freedom from unwanted interference and noise, whether this originates within the receiver circuit or from external sources;
- **linearity** – which implies an absence of any distortion of the signal during the transmission/reception process.

These requirements for receiver performance will be discussed later under the heading of receiver design. However, the quality of the signal heard by the listener depends very largely on the nature of the signal present at the receiver. This depends, in the first place, on the transmitter and the transmission techniques employed.

In normal public service broadcasting – where it is required that the signal shall be received, more or less uniformly, throughout the entire service area – the transmitter aerial is designed so that it has a uniform, 360°, dispersal pattern. Also the horizontal shape of the transmission 'lobe' (the conventional pictorial representation of relative signal strength, as a function of the angle) is as shown in Fig. 2.1.

The influence of ground attenuation, and the curvature of the earth's surface, mean that in this type of transmission pattern the signal strength, gets progressively weaker as the height above ground level of the receiving aerial gets less, except in the immediate neighbourhood of the transmitter. There are a few exceptions to this rule, as will be shown later, but it is generally true, and implies that the higher the receiver aerial can be placed, in general the better.

THE INFLUENCE OF THE IONOSPHERE

The biggest modifying influence on the way the signal reaches the receiver is the presence of a reflecting – or, more strictly, refracting – ionised

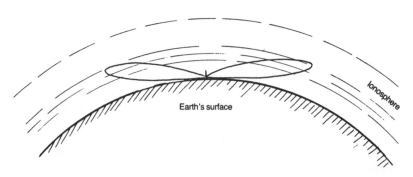

Fig. 2.1   *Typical transmitter aerial lobe pattern.*

band of gases in the outer regions of the earth's atmosphere. This is called the ionosphere and is due to the incidence of a massive bombardment of energetic particles on the outer layers of the atmosphere, together with ultra-violet and other electromagnetic radiation, mainly from the sun.

This has the general pattern shown in Fig. 2.2 if plotted as a measure of electron density against height from the surface. Because it is dependent on radiation from the sun, its strength and height will depend on whether the earth is exposed to the sun's radiation (daytime) or protected by its shadow (night).

As the predominant gases in the earth's atmosphere are oxygen and nitrogen, with hydrogen in the upper reaches, and as these gases tend to separate somewhat according to their relative densities, there are three effective layers in the ionosphere. These are the 'D' (lowest) layer, which contains ionised oxygen/ozone; the 'E' layer, richer in ionised nitrogen and nitrogen compounds; and the 'F' layer (highest), which largely consists of ionised hydrogen.

Since the density of the gases in the lower layers is greater, there is a much greater probability that the ions will recombine and disappear, in the absence of any sustaining radiation. This occurs as the result of normal collisions of the particles within the gas, so both the 'D' and the 'E' layers tend to fade away as night falls, leaving only the more rarified 'F' layer. Because of the lower gas pressure, molecular collisions will occur more rarely in the 'F' layer, but here the region of maximum electron density tends to vary in mean height above ground level.

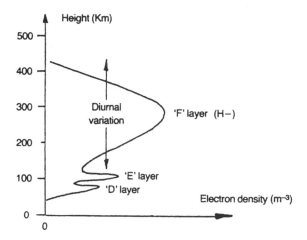

Fig. 2.2 *The electron density in the ionosphere.*

Critical frequency

The way in which radio waves are refracted by the ionosphere, shown schematically in Fig. 2.3, is strongly dependent on their frequency, with a 'critical frequency' ('F_c') dependent on electron density, per cubic metre, according to the equation

$$F_c = 9\sqrt{N_{max}}$$

where N_{max} is the maximum density of electrons/cubic metre within the layer. Also, the penetration of the ionosphere by radio waves increases as the frequency is increased. So certain frequency bands will tend to be refracted back towards the earth's surface at different heights, giving different transmitter to receiver distances for optimum reception, as shown in Fig. 2.4, while some will not be refracted enough, and will continue on into outer space.

The dependence of radio transmission on ionosphere conditions, which, in turn depends on time of day, time of year, geographical latitude, and 'sun spot' activity, has led to the term 'MUF' or maximum usable frequency, for such transmissions.

Also, because of the way in which different parts of the radio frequency spectrum are affected differently by the possibility of ionospheric refraction, the frequency spectrum is classified as shown in Table 2.1. In this VLF and LF signals are strongly refracted by the 'D' layer, when present, MF signals by the 'E' and 'F' layers, and HF signals only by the 'F' layer, or not at all.

Additionally, the associated wavelengths of the transmissions (from 100 000–1000 m in the case of the VLF and LF signals, are so long that the earth's surface appears to be smooth, and there is a substantial

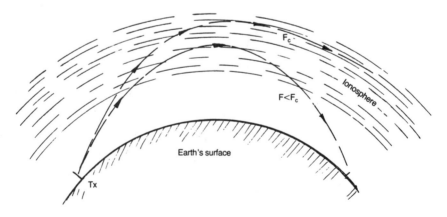

Fig. 2.3 *The refraction of radio waves by the ionosphere.*

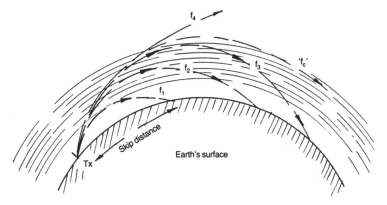

Fig. 2.4 *The influence of frequency on the optimum transmitter to receiver distance — the 'skip distance'.*

Table 2.1 Classification of radio frequency spectrum

VLF	3–30	kHz
LF	30–300	kHz
MF	300–3000	kHz
HF	3–30	MHz
VHF	30–300	MHz
UHF	300–3000	MHz
SHF	3–30	GHz

reflected 'ground wave' which combines with the direct and reflected radiation to form what is known as a 'space wave'. This space wave is capable of propagation over very long distances, especially during daylight hours when the 'D' and 'E' layers are strong.

VHF/SHF effects

For the VHF to SHF regions, different effects come into play, with very heavy attenuation of the transmitted signals, beyond direct line-of-sight paths, due to the intrusions of various things which will absorb the signal, such as trees, houses, and undulations in the terrain. However, temperature inversion layers in the earth's atmosphere, and horizontal striations in atmospheric humidity, also provide mechanisms, especially at the higher end of the frequency spectrum, where the line of sight paths may be extended somewhat to follow the curvature of the earth.

Only certain parts of the available radio frequency (RF) spectrum are allocated for existing or future commercial broadcast use. These have been subclassified as shown in Table 2.2, with the terms 'Band 1' to 'Band 6' being employed to refer to those regions used for TV and FM radio.

The internationally agreed FM channel allocations ranged, originally, from Channel 1 at 87.2−87.4 MHz, to Channel 60, at 104.9−105.1 MHz, based on 300 kHz centre-channel frequency spacings. However, this allocation did not take into account the enormous future proliferation of local transmitting stations, and the centre-channel spacings are now located at 100 kHz intervals.

Depending on transmitted power, transmitters will usually only be operated at the same frequency where they are located at sites which are remote from each other. Although this range of operating frequencies somewhat overlaps the lower end of 'Band 2', all of the UK operating frequencies stay within this band.

Table 2.2   Radio broadcast band allocations

Wavelength	Allocation	Band
LW	150−285 kHz	
MW	525−1605 kHz	
SW	5.95−6.2 MHz	49 M
	7.1−7.3 MHz	40 M
	9.5−9.775 MHz	30 M
	11.7−11.975 MHz	25 M
	15.1−15.45 MHz	19 M
	17.7−17.9 MHz	16 M
	21.45−21.75 MHz	13 M
	25.5−26.1 MHz	11 M

Note National broadcasting authorities may often overspill these frequency limits.

Band	Wavelength allocation
I	41−68 MHz
II	87.5−108 MHz
III	174−223 MHz
IV	470−585 MHz
V	610−960 MHz
VI	11.7−12.5 GHz

WHY VHF TRANSMISSIONS?

In the early days of broadcasting, when the only reliable long to medium distance transmissions were thought to be those in the LF-MF regions of the radio spectrum, (and, indeed, the HF bands were handed over to amateur use because they were thought to be of only limited usefulness), broadcast transmitters were few and far between. Also the broadcasting authorities did not aspire to a universal coverage of their national audiences. Under these circumstances, individual transmitters could broadcast a high quality, full frequency range signal, without problems due to adjacent channel interference.

However, with the growth of the aspirations of the broadcasting authorities, and the expectations of the listening public, there has been an enormous proliferation of radio transmitters. There are now some 440 of these in the UK alone, on LW, MW and Band 2 VHF allocations, and this ignores the additional 2400 odd separate national and local TV transmissions.

If all the radio broadcast services were to be accommodated within the UK's wavelength allocations on the LW and MW bands, the congestion would be intolerable. Large numbers would have to share the same frequencies, with chaotic mutual interference under any reception conditions in which any one became receivable in an adjacent reception area.

Frequency choice

The decision was therefore forced that the choice of frequencies for all major broadcast services, with the exception of pre-existing stations, must be such that the area covered was largely line-of-sight, and unaffected by whether it was day or night.

It is true that there are rare occasions when, even on Band 2 VHF, there are unexpected long-distance intrusions of transmissions. This is due to the occasional formation of an intense ionisation band in the lower regions of the ionosphere, known as 'sporadic E'. Its occurrence is unpredictable and the reasons for its occurrences are unknown, although it has been attributed to excess 'sun spot' activity, or to local thermal ionisation, as a result of the shearing action of high velocity upper atmosphere winds.

In the case of the LW and MW broadcasts, international agreements aimed at reducing the extent of adjacent channel interference have continually restricted the bandwidth allocations available to the broadcasting authorities. For MW at present, and for LW broadcasts as from 1 February 1988 – ironically the centenary of Hertz's original experiment – the channel separation is 9 kHz and the consequent maximum transmitted audio bandwidth is 4.5 kHz.

In practice, not all broadcasting authorities conform to this restraint, and even those that do interpret it as 'flat from 30 Hz to 4.5 kHz and −50 dB at 5.5 kHz' or more leniently, as 'flat to 5 kHz, with a more gentle roll-off to 9 kHz'. However, by comparison with the earlier accepted standards for MW AM broadcasting, of 30 Hz−12 kHz, ±1 dB, and with less than 1% THD at 1 kHz, 80% modulation, the current AM standards are poor.

One also may suspect that the relatively low standards of transmission quality practicable with LF/MF AM broadcasting encourages some broadcasting authorities to relax their standards in this field, and engage in other, quality degrading, practices aimed at lessening their electricity bills.

AM or FM?

Having seen that there is little scope for high quality AM transmissions on the existing LW and MW broadcast bands, and that VHF line-of-sight transmissions are the only ones offering adequate bandwidth and freedom from adjacent channel interference, the question remains as to what style of modulation should be adopted to combine the programme signal with the RF carrier.

Modulation systems

Two basic choices exist, that of modulating the amplitude of the RF carrier, (AM), or of modulating its frequency, (FM), as shown in Fig. 2.5. The technical advantages of FM are substantial, and these were confirmed in a practical field trial in the early 1950s carried out by the BBC, in which the majority of the experimental listening panel expressed a clear preference for the FM system.

The relative qualities of the two systems can be summarised as follows. AM is:

- the simplest type of receiver to construct
- not usually subject to low distortion in the recovered signal
- prone to impulse-type (e.g. car ignition) interference and to 'atmospherics'
- possibly subject to 'fading'
- prone to adjacent channel or co-channel interference
- affected by tuning and by tuned circuit characteristics in its frequency response.

FM:

- requires more complex and sophisticated receiver circuitry

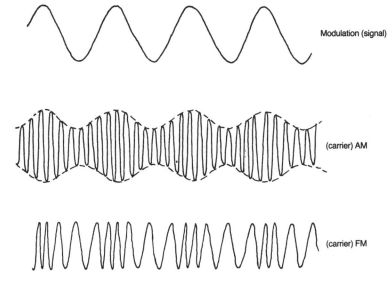

Fig. 2.5 *Carrier modulation systems.*

- can give very low signal distortion
- ensures all received signals will be at the same apparent strength
- is immune to fading
- is immune to adjacent channel and co-channel interference, provided that the intruding signals are less strong.
- has, potentially, a flat frequency response, unaffected by tuning or tuned circuit characteristics
- will reject AM and impulse-type interference
- makes more efficient use of available transmitter power, and gives, in consequence, a larger usable reception area.

On the debit side, the transmission of an FM signal requires a much greater bandwidth than the equivalent AM one, in a ratio of about 6:1. The bandwidth requirements may be calculated approximately from the formula

$$B_n = 2M + 2DK$$

where B_n is the necessary bandwidth, M is the maximum modulation frequency in Hz, D is the peak deviation in Hz, and K is an operational constant (= 1 for a mono signal).

However, lack of space within the band is not a significant problem at VHF, owing to the relatively restricted geographical area covered by any transmitter. So Band 2 VHF/FM has become a worldwide choice for high quality transmissions, where the limitations are imposed more by the distribution method employed to feed programme signals to the transmitter, and by the requirements of the stereo encoding/decoding process than by the transmitter or receiver.

FM BROADCAST STANDARDS

It is internationally agreed that the FM frequency deviation will be 75 kHz, for 100% modulation. The stereo signal will be encoded, where present, by the Zenith-GE 38 kHz sub-carrier system, using a 19 kHz ± 2 Hz pilot tone, whose phase stability is better than 3° with reference to the 38 kHz sub-carrier. Any residual 38 kHz sub-carrier signal present in the composite stereo output shall be less than 1%.

Local agreements, in Europe, specify a 50 μs transmission pre-emphasis. In the USA and Japan, the agreed pre-emphasis time constant is 75 μs. This gives a slightly better receiver S/N ratio, but a somewhat greater proneness to overload, with necessary clipping, at high audio frequencies.

STEREO ENCODING/DECODING

One of the major attractions of the FM broadcasting system is that it allows the transmission of a high quality stereo signal, without significant degradation of audio quality – although there is some worsening of S/N ratio. For this purpose the permitted transmitter bandwidth is 240 kHz, which allows an audio bandwidth, on a stereo signal, of 30 Hz−15 kHz, at 90% modulation levels. Lower modulation levels would allow a more extended high-frequency audio bandwidth, up to the 'zero transmission above 18.5 kHz' limit imposed by the stereo encoding system.

It is not known that any FM broadcasting systems significantly exceed the 30 Hz−15 kHz audio bandwidth levels.

Because the 19 kHz stereo pilot tone is limited to 10% peak modulation, it does not cause these agreed bandwidth requirements to be exceeded.

GE/ZENITH 'PILOT TONE' SYSTEM

This operates in a manner which produces a high-quality 'mono' signal in a receiver not adapted to decode the stereo information, by the transmission of a composite signal of the kind represented in Fig. 2.6. In this the combined left-hand channel and right-hand channel (L+R) − mono − signal is transmitted normally in the 30 Hz−15 kHz part of the spectrum, with a maximum modulation of 90% of the permitted 75 kHz deviation.

An additional signal, composed of the difference signal between these

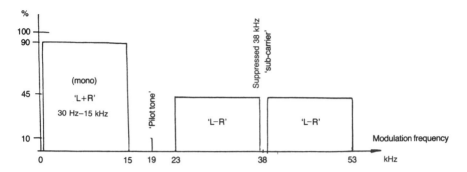

Fig. 2.6 *The Zenith-GE 'pilot tone' stereophonic system.*

channels, (L−R), is then added as a modulation on a suppressed 38 kHz sub-carrier. So that the total modulation energy will be the same, after demodulation, the modulation depth of the combined (L−R) signal is held to 45% of the permitted maximum excursion. This gives a peak deviation for the transmitted carrier which is the same as that for the 'mono' channel.

Decoding

This stereo signal can be decoded in two separate ways, as shown in Figs 2.7 and 2.8. In essence, both of these operate by the addition of the two (L+R) and (L−R) signals to give LH channel information only, and the subtraction of these to give the RH channel information only.

In the circuit of Fig. 2.7, this process is carried out by recovering the separate signals, and then using a matrix circuit to add or subtract them. In the circuit of Fig. 2.8, an equivalent process is done by sequentially sampling the composite signal, using the regenerated 38 kHz sub-carrier to operate a switching mechanism.

Advocates of the matrix addition method of Fig. 2.7 have claimed that this allows a better decoder signal-to-noise (S/N) ratio than that cf the sampling system. This is only true if the input bandwidth of the sampling system is sufficiently large to allow noise signals centred on the harmonics of the switching frequency also to be commutated down into the audio spectrum.

Provided that adequate input filtration is employed in both cases there is no operational advantage to either method. Because the system shown in Fig. 2.8 is more easily incorporated within an integrated circuit, it is very much the preferred method in contemporary receivers.

In both cases it is essential that the relative phase of the regenerated

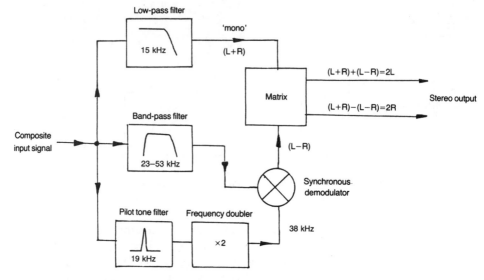

Fig. 2.7   *Matrix addition type of stereo decoder.*

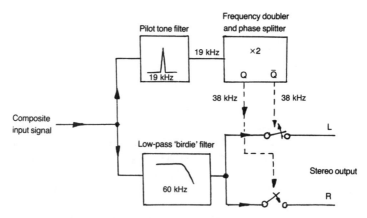

Fig. 2.8   *Synchronous switching stereo decoder.*

38 kHz sub-carrier is accurately related to the composite incoming signal. Errors in this respect will degrade the 35–40 dB (maximum) channel separation expected with this system.

Because the line bandwidth or sampling frequency of the studio to transmitter programme link may not be compatible with the stereo encoded signal, this is normally encoded on site, at the transmitter, from the received LH and RH channel information. This is done by the method shown in Fig. 2.9.

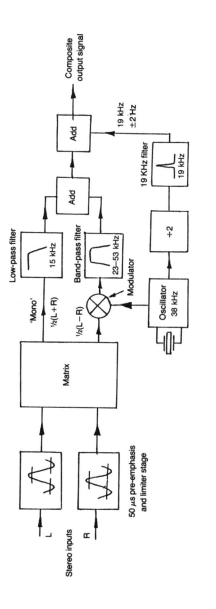

Fig. 2.9 *Zenith-GE stereophonic encoding system.*

In this, the incoming LH and RH channel signals (pre-emphasis and peak signal level limitation will generally have been carried out before this stage) are combined together in a suitable matrix – the simple double transformer of Fig. 2.10 would serve – and fed to addition networks in the manner shown.

In the case of the (L−R) signal, however, it is first converted into a modulated signal based on a 38 kHz sub-carrier, derived from a stable crystal-controlled oscillator, using a balanced modulator to ensure suppression of the residual 38 kHz sub-carrier frequency. It is then filtered to remove any audio frequency components before addition to the (L+R) channel.

Finally, the 38 kHz sub-carrier signal is divided, filtered and phase corrected to give a small amplitude, 19 kHz sine-wave pilot tone which can be added to the composite signal before it is broadcast.

The reason for the greater background 'hiss' associated with the stereo than with the mono signal is that wide-band noise increases as the square-root of the audio bandwidth. In the case of a mono signal this bandwidth is 30 Hz−15 kHz. In the case of a Zenith-GE encoded stereo signal it will be at least 30 Hz−53 kHz, even if adequate filtering has been used to remove spurious noise components based on the 38 kHz sub-carrier harmonics.

THE BBC PCM PROGRAMME DISTRIBUTION SYSTEM

In earlier times, with relatively few, high-power, AM broadcast transmitters, it was practicable to use high-quality telephone lines as the means of routing the programme material from the studio to the transmitter. This might even, in some cases, be in the same building.

However, with the growth of a network of FM transmitters serving local areas it became necessary to devise a method which would allow consistently high audio quality programme material to be sent over long distances without degradation. This problem became more acute with the spread of stereo broadcasting, where any time delay in one channel with respect to the other would cause a shift in the stereo image location.

The method adopted by the BBC, and designed to take advantage of the existing 6.5 MHz bandwidth TV signal transmission network, was to convert the signal into digital form, in a manner which was closely analogous to that adopted by Philips for its compact disc system. However, in the case of the BBC a rather lower performance standard was adopted. The compact disc uses a 44.1 kHz sampling rate, a 16-bit (65536 step) sampling resolution, and an audio bandwidth of 20 Hz−20 kHz, (±0.2 dB), while the BBC system uses a 32 kHz sampling rate, a 13-bit (8192 step) resolution and a 50 Hz−14.5 kHz bandwidth, (± 0.2 dB).

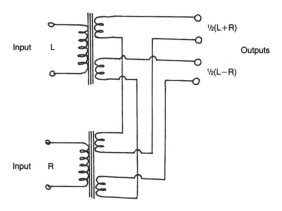

Fig. 2.10 *Simple matrixing method.*

The BBC system offers a CCIR weighted S/N ratio of 57 dB, and a non-linear distortion figure of 0.1% ref. full modulation at 1 kHz.

As in all digital systems, the prominence of the 'staircase type' step discontinuity becomes greater at low signal levels and higher audio frequencies, giving rise to a background noise associated with the signal, known as 'quantisation noise'.

Devotees of hi-fi tend to refer to these quantisation effects as 'granularity'. Nevertheless, in spite of the relatively low standards, in the hi-fi context, adopted for the BBC PCM transmission links, there has been relatively little criticism of the sound quality of the broadcasts.

The encoding and decoding systems used are shown in Figs 2.11 and 2.12.

Encoding

The operation of the PCM encoder, shown schematically in Fig. 2.11, may be explained by consideration of the operation of a single input channel. In this, the incoming signal is filtered, using a filter with a very steep attenuation rate, to remove all signal components above 15 kHz.

This is necessary since with a 32 kHz sampling rate, any audio components above half the sampling rate (say at 17 kHz) would be resolved identically to those at an equal frequency separation below this rate, (e.g. at 15 kHz), and this could cause severe problems both due to spurious signals, and to intermodulation effects between these signals. This problem is known as 'aliasing', and the filters are known as 'anti-aliasing' filters.

Because the quality of digitally encoded waveforms deteriorates as the signal amplitude gets smaller, it is important to use the largest practicable signal levels (in which the staircase-type granularity will be as small a

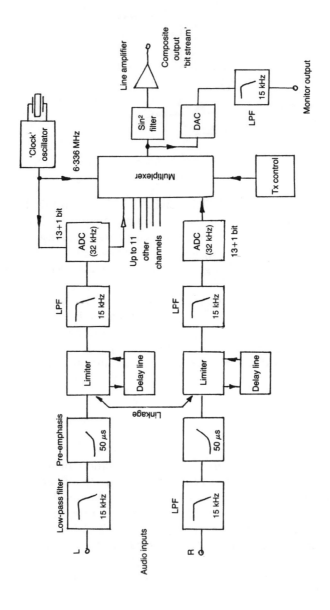

Fig. 2.11 *The BBC 13-channel pulse code modulation (PCM) encoder system.*

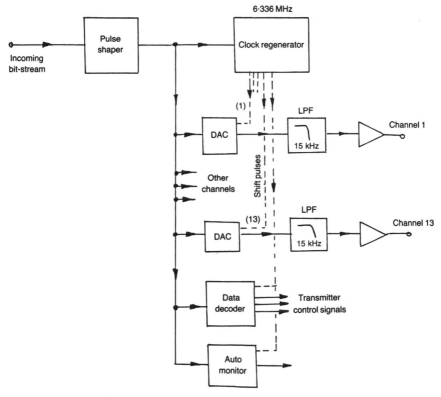

Fig. 2.12   *The BBC PCM decoding system.*

proportion of the signal as possible). It is also important to ensure that the analogue-to-digital encoder is not overloaded. For this reason delay-line type limiters are used, to delay the signal long enough for its amplitude to be assessed and appropriate amplitude limitation to be applied.

In order to avoid hard peak clipping which is audibly displeasing, the limiters operate by progressive reduction of the stage gain, and have an output limit of +2 dB above the nominal peak programme level. Carrying out the limiting at this stage avoids the need for a further limiter at the transmitter. Pre-emphasis is also added before the limiter stage. The effect of this pre-emphasis on the frequency response is shown in Fig. 2.13.

An interesting feature of the limiter stages is that those for each stereo line pair are linked, so that if the peak level is exceeded on either, both channels are limited. This avoids any disturbance in stereo image location which might arise through a sudden amplitude difference between channels.

The AF bandwidth limited signal from the peak limiter is then converted

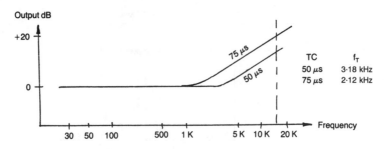

Fig. 2.13 *HF pre-emphasis characteristics.*

into digital form by a clocked 'double ramp' analogue-to-digital converter, and fed, along with the bit streams from up to 12 other channels, to a time-domain multiplexing circuit. The output from this is fed through a filter, to limit the associated RF bandwidth, to an output line amplifier.

The output pulses are of sine-square form, with a half amplitude duration of 158 ns, and 158 ns apart. Since the main harmonic component of these pulses is the second, they have a negligible energy distribution above 6.336 MHz, which allows the complete composite digital signal to be handled by a channel designed to carry a 625-line TV programme signal.

The output of the multiplexer is automatically sampled, sequentially, at 256 ms per programme channel, in order to give an automatic warning of any system fault. It can also be monitored as an audio output signal reference.

To increase the immunity of the digitally encoded signal to noise or transmission line disturbances, a parity bit is added to each preceding 13-bit word. This contains data which provide an automatic check on the five most significant digits in the preceding bit group. If an error is detected, the faulty word is rejected and the preceding 13-bit word is substituted. If the error persists, the signal is muted until a satisfactory bit sequence is restored.

Effectively, therefore, each signal channel comprises 14 bits of information. With a sampling rate of 32 kHz, the 6.336 Mbits/s bandwidth allows a group of 198 bits in each sample period. This is built from 13 multiplexed 14-bit channels (= 182 bits) and 16 'spare' bits. Of these, 11 are used to control the multiplexing matrix, and four are used to carry transmitter remote control instructions.

Decoding

The decoding system is shown in Fig. 2.12. In this the incoming bit stream is cleaned up, and restored to a sequence with definite '0' and '1' states. The data stream is then used to regenerate the original 6336 kHz clock signal, and a counter system is used to provide a sequence of shift pulses fed, in turn, to 13 separate digital-to-analogue (D-A) converters, which recreate the original input channels, in analogue form.

A separate data decoder is used to extract the transmitter control information, and, as before, an automatic sampling monitor system is used to check the correct operation of each programme channel.

The availability of spare channels can be employed to carry control information on a digitally controlled contrast compression/expansion (Compander) system (such as the BBC NICAM 3 arrangement). This could be used either to improve the performance, in respect of the degradation of small signals, using the same number of sample steps, or to obtain a similar performance with a less good digital resolution (and less transmission bandwidth per channel).

The standards employed for TV sound links, using a similar type of PCM studio-transmitter link, are:

- AF bandwidth, 50 Hz−13.5 kHz (± 0.7 dB ref. 1 kHz)
- S/N ratio, 53 dB CCIR weighted
- non-linear distortion and quantisation defects, 0.25% (ref. 1 kHz and max. modulation).

SUPPLEMENTARY BROADCAST SIGNALS

It has been suggested that data signals could be added to the FM or TV programme signal, and this possibility is being explored by some European broadcasting authorities. The purpose of this would be to provide signal channel identification, and perhaps also to allow automatic channel tuning. The system proposed for this would use an additional low level sub-carrier, for example, at 57 kHz. In certain regions of Germany additional transmissions based on this sub-carrier frequency are used to carry the VWF (road/traffic conditions) information broadcasts.

In the USA, 'Storecast' or 'Subsidiary Communication Authorisation' (SCA) signals may be added to FM broadcasts, to provide a low-quality 'background music' programme for subscribers to this scheme. This operates on a 67 kHz sub-carrier, with a max. modulation level of 10%.

ALTERNATIVE TRANSMISSION METHODS

Apart from the commercial (entertainment and news) broadcasts on the LW, MW and specified short wave bands shown in Table 2.2, where the transmission techniques are exclusively double-sideband amplitude modulated type, and in Band 2 (VHF) where the broadcasts are exclusively frequency modulated, there are certain police, taxi and other public utility broadcasts on Band 2 which are amplitude-modulated.

It is planned that these other Band 2 broadcasts will eventually be moved to another part of the frequency spectrum, so that the whole of Band 2 can be available for FM radio transmissions.

However, there are also specific bands of frequencies, mainly in the HF/VHF parts of the spectrum, which have been allocated specifically for amateur use. These are shown in Table 2.3.

In these amateur bands, other types of transmitted signal modulation may be employed. These are narrow-bandwidth FM (NBFM), phase modulation (PM), and single-sideband suppressed carrier AM (SSB).

The principal characteristics of these are that NBFM is restricted to a total deviation of ± 5 kHz, and a typical maximum modulation frequency of 3 kHz, limited by a steep-cut AF filter. This leads to a low modulation index (the ratio between the carrier deviation and the modulating frequency), which leads to poor audio quality and difficulties in reception (demodulation) without significant distortion. The typical (minimum) bandwidth requirement for NBFM is 13 kHz.

Phase modulation (PM), shares many of the characteristics of NBFM, except that in PM, the phase deviation of the signal is dependent both upon the amplitude and the frequency of the modulating signal, whereas in FM the deviation is purely dependent on programme signal amplitude.

Both of these modulation techniques require fairly complex and well designed receivers, if good reception and S/N ratio is to be obtained.

SSB BROADCASTING

The final system, that of suppressed carrier single-sideband transmission (SSB), is very popular among the amateur radio transmitting fraternity, and will be found on all of the amateur bands.

Table 2.3 Amateur frequency band allocations

Band	MHz
80 M	3.5−3.725
40 M	7.0−7.15
20 M	14.0−14.35
15 M	21.0−21.45
10 M	28.0−29.7
6 M	50.0−54.0
2 M	144.0−148.0

This relies on the fact that the transmitted frequency spectrum of an AM carrier contains two identical, mirror-image, groups of sidebands below (lower sideband or LSB) and above (upper sideband or USB) the carrier, as shown in Fig. 2.14. The carrier itself conveys no signal and serves merely as the reference frequency for the sum and difference components which together reconstitute the original modulation.

If the receiver is designed so that a stable carrier frequency can be reintroduced, both the carrier and one of the sidebands can be removed entirely, without loss of signal intelligibility. This allows a very much larger proportion of the transmitter power to be used for conveying signal information, with a consequent improvement in range and intelligibility. It also allows more signals to be accomodated within the restricted bandwidth allocation, and gives better results with highly selective radio receivers than would otherwise have been possible. The method employed is shown schematically in Fig. 2.15.

There is a convention among amateurs that SSB transmissions up to 7 MHz shall employ LSB modulation, and those above this frequency shall use USB. This technique is not likely to be of interest to those concerned with good audio quality, but its adoption for MW radio broadcast reception is proposed from time to time, as a partial solution to the poor audio bandwidth possible with double sideband (DSB) transmission at contemporary 9 kHz carrier frequency separations.

The reason for this is that the DSB transmission system used at present only allows a 4.5 kHz AF bandwidth, whereas SSB would allow a full 9 kHz audio pass-band. On the other hand, even a small frequency drift of the reinserted carrier frequency can transpose all the programme frequency components upwards or downwards by a step frequency interval, and this can make music sound quite awful; so any practical method would have to rely on the extraction and amplification of the residue of the original carrier, or on some equally reliable carrier frequency regeneration technique.

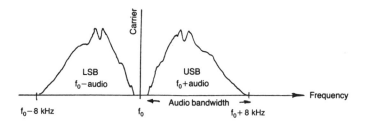

Fig. 2.14 *Typical AM bandwidth spectrum for double-sideband transmission.*

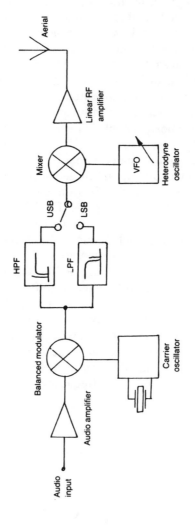

Fig. 2.15 *Single sideband (USB or LSB) transmitter system.*

RADIO RECEIVER DESIGN

The basic requirements for a satisfactory radio receiver were listed above. Although the relative importance of these qualities will vary somewhat from one type of receiver to another — say, as between an AM or an FM receiver — the preferred qualities of radio receivers are, in general, sufficiently similar for them to be considered as a whole, with some detail amendments to their specifications where system differences warrant this.

Selectivity

An ideal selectivity pattern for a radio receiver would be one which had a constant high degree of sensitivity at the desired operating frequency, and at a sufficient bandwidth on either side of this to accommodate transmitter sidebands, but with zero sensitivity at all frequencies other than this.

A number of techniques have been used to try to approach this ideal performance characteristic. Of these, the most common is that based on the inductor-capacitor (LC) parallel tuned circuit illustrated in Fig. 2.16(a).

Tuned circuit characteristics

For a given circulating RF current, induced into it from some external source, the AC potential appearing across an L–C tuned circuit reaches a peak value at a frequency (F_o) given by the equation

$$F_o = 1/(2\pi\sqrt{LC})$$

Customarily the group of terms $2\pi F$ are lumped together and represented

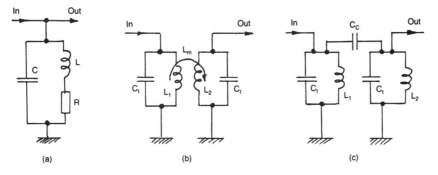

(a) (b) (c)

Fig. 2.16 *Single (a) and bandpass (b) (c) tuned circuits.*

by the symbol ω, so that the peak output, or resonant frequency would be represented by

$$\omega_o = \frac{1}{\sqrt{LC}}$$

Inevitably there will be electrical energy losses within the tuned circuit, which will degrade its performance, and these are usually grouped together as a notional resistance, r, appearing in series with the coil.

The performance of such tuned circuits, at resonance, is quantified by reference to the circuit magnification factor or quality factor, referred to as Q. For any given L–C tuned circuit this can be calculated from

$$Q = \frac{\omega_o L}{r} \text{ or } \frac{1}{\omega_o C r}$$

Since, at resonance, (ω_o), $\omega_o = 1/\sqrt{LC}$, the further equation

$$Q = \frac{1}{r}\sqrt{\frac{L}{C}}$$

can be derived. This shows that the Q improves as the equivalent loss resistance decreases, and as the ratio of L to C increases.

Typical tuned circuit Q values will lie between 50 and 200, unless the Q value has been deliberately degraded, in the interests of a wider RF signal pass-band, usually by the addition of a further resistor, R, in parallel with the tuned circuit.

The type of selectivity offered by such a single tuned circuit is shown, for various Q values, in Fig. 2.17(a). The actual discrimination against frequencies other than the desired one is clearly not very good, and can be calculated from the formula

$$\delta F = F_o/2Q$$

where δF is the 'half-power' bandwidth.

One of the snags with single tuned circuits, which is exaggerated if a number of such tuned circuits are arranged in cascade to improve the selectivity, is that the drop in output voltage from such a system, as the frequencies differ from that of resonance, causes a very rapid attenuation of higher audio frequencies in any AM type receiver, in which such a tuning arrangement was employed, as shown in curve 'a' of Fig. 2.18.

Clearly, this loss of higher audio frequencies is quite unacceptable, and the solution adopted for AM receivers is to use pairs of tuned circuits, coupled together by mutual inductance, L_m, or, in the case of separated circuits, by a coupling capacitor, C_c.

This leads to the type of flat-topped frequency response curve shown in

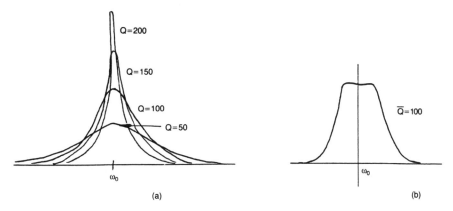

Fig. 2.17 *Response curves of single (a) and bandpass (b) tuned circuits.*

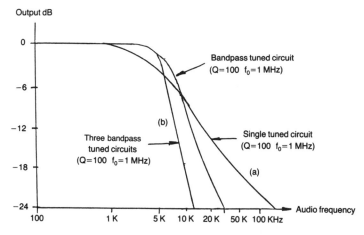

Fig. 2.18 *Effect on AM receiver AF response of tuned circuit selectivity characteristics.*

Fig. 2.17(b), when critical coupling is employed, when the coupling factor

$$k = 1/\sqrt{Q_1 Q_2}$$

For a mutually inductive coupled tuned circuit, this approximates to

$$L_m = \bar{L}/Q$$

and for a capacitatively coupled bandpass tuned circuit.

$$C_c = \bar{C}_t/Q$$

where L_m is the required mutual inductance, $\bar{L}$ is the mean inductance of the coils, $\bar{C}_t$ is the mean value of the tuning capacitors, and C_c is the required coupling capacitance.

In the case of a critically-coupled bandpass tuned circuit, the selectivity is greater than that given by a single tuned circuit, in spite of the flat-topped frequency response. Taking the case of $F_o = 1$ MHz, and $Q = 100$, for both single- and double-tuned circuits the attenuation will be -6 dB at 10 kHz off tune. At 20 kHz off tune, the attenuation will be -12 dB and -18 dB respectively, and at 30 kHz off tune it will be -14 dB and -24 dB, for these two cases.

Because of the flat-topped frequency response possible with bandpass-coupled circuits, it is feasible to put these in cascade in a receiver design without incurring audio HF loss penalties at any point up to the beginning of the attenuation 'skirt' of the tuned circuit. For example, three such groups of 1 MHz bandpass-tuned circuits, with a circuit Q of 100, would give the following selectivity characteristics.

- 5 kHz = 0 dB
- 10 kHz = -18 dB
- 20 kHz = -54 dB
- 30 kHz = -72 dB

as illustrated in curve 'b' in Fig. 2.18. Further, since the bandwidth is proportional to frequency, and inversely proportional to Q, the selectivity characteristics for any other frequency or Q value can be derived from these data by extrapolation.

Obviously this type of selectivity curve falls short of the ideal, especially in respect of the allowable audio bandwidth, but selectivity characteristics of this kind have been the mainstay of the bulk of AM radio receivers over the past sixty or more years. Indeed, because this represents an ideal case for an optimally designed and carefully aligned receiver, many commercial systems would be less good even than this.

For FM receivers, where a pass bandwidth of 220–250 kHz is required, only very low Q tuned circuits are usable, even at the 10.7 MHz frequency at which most of the RF amplification is obtained. So an alternative technique is employed, to provide at least part of the required adjacent channel selectivity. This is the surface acoustic wave filter.

The surface acoustic wave (SAW) filter

This device, illustrated in schematic form in Fig. 2.19, utilises the fact that it is possible to generate a mechanical wave pattern on the surface of a piece of piezo-electric material (one in which the mechanical dimensions will change under the influence of an applied electrostatic field), by

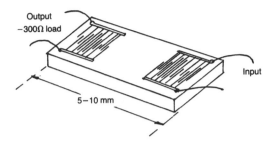

Fig. 2.19 *Construction of surface acoustic wave (SAW) filter.*

applying an AC voltage between two electrically conductive strips laid on the surface of the material.

Since the converse effect also applies — that a voltage would be induced between these two conducting strips if a surface wave in a piezo-electric material were to pass under them — this provides a means for fabricating an electro-mechanical filter, whose characteristics can be tailored by the number, separation, and relative lengths of the conducting strips.

These SAW filters are often referred to as IDTs (inter-digital transducers), from the interlocking finger pattern of the metallising, or simply as ceramic filters, since some of the earlier low-cost devices of this type were made from piezo-electric ceramics of the lead zirconate-titanate (PZT) type. Nowadays they are fabricated from thin, highly polished strips of lithium niobate, (LiNbO), bismuth germanium oxide, (BiGeO), or quartz on which a very precise pattern of metallising has been applied by the same photo-lithographic techniques employed in the manufacture of integrated circuits.

Two basic types of SAW filter are used, of which the most common is the transversal type. Here a surface wave is launched along the surface from a transmitter group of digits to a receiver group. This kind is normally used for bandpass applications. The other form is the resonant type, in which the surface electrode pattern is employed to generate a standing-wave effect.

Because the typical propagation velocity of such surface waves is of the order of 3000 m/s, practical physical dimensions for the SAW slices and conductor patterns allow a useful frequency range without the need for excessive physical dimensions or impracticably precise electrode patterns. Moreover, since the wave only penetrates a wavelength or so beneath the surface, the rear of the slice can be cemented onto a rigid substrate to improve the ruggedness of the element.

In the transversal or bandpass type of filter, the practicable range of operating frequencies is, at present, from a few MHz to 1–2 GHz, with minimum bandwidths of around 100 kHz. The resonant type of SAW device can operate from a few hundred kHz to a similar maximum frequency, and offers a useful alternative to the conventional quartz crystal (bulk resonator) for the higher part of the frequency range, where bulk resonator systems need to rely on oscillator overtones.

The type of performance characteristic offered by a bandpass SAW device, operating at a centre frequency of 10.7 MHz, is shown in Fig. 2.20. An important quality of wide pass-band SAW filters is that the phase and attenuation characteristics can be separately optimised by manipulating the geometry of the pattern of the conducting elements deposited on the surface.

This is of great value in obtaining high audio quality from FM systems, as will be seen below, and differs, in this respect, from tuned circuits or other types of filter where the relative phase-angle of the transmitted signal is directly related to the rate of change of transmission as a function of frequency, in proximity to the frequency at which the relative phase-angle is being measured.

However, there are snags. Good quality phase linear SAW filters are expensive, and there is a relatively high insertion loss, typically in the range of −15 to −25 dB, which requires additional compensatory amplification. On the other hand, they are physically small and do not suffer, as do coils, from unwanted mutual induction effects. The characteristic impedance of such filters is, typically, 300−400 ohms, and changes in the source and load impedances can have big effects on the transmission curves.

THE SUPERHET SYSTEM

It will have been appreciated from the above that any multiple tuned circuit, or SAW filter, system chosen to give good selectivity will be optimised only for one frequency. To tune a group of critically coupled

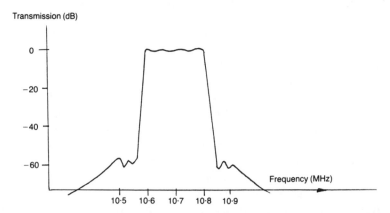

Fig. 2.20   *Transmission characteristics of typical 10.7 MHz SAW filter.*

bandpass tuned circuits simultaneously to cover a required receiver frequency range would be very difficult. To tune one or more SAW filters simultaneously would simply not be possible at all, except, perhaps, over an exceedingly limited frequency range.

A practical solution to this problem was offered in 1918 by Major Armstrong of the US Army, in the form of the supersonic heterodyne or superhet receiver system, shown in Fig. 2.21.

In this, the incoming radio signal is combined with an adjustable frequency local oscillator signal in some element having a non-linear input/output transfer characteristic, (ideally one having a square-law slope). This stage is known as the frequency changer or mixer or, sometimes, and inappropriately, as the first detector.

This mixture of the two (input and LO) signals gives rise to additional sum and difference frequency outputs, and if the local oscillator frequency is chosen correctly, one or other of these can be made to coincide with the fixed intermediate frequency (usually known as the IF), at which the bulk of the amplification will occur.

The advantages of this arrangement, in allowing the designer to tailor his selectivity characteristics without regard to the input frequency, are enormous, and virtually all commercial radio receivers employ one or other of the possible permutations of this system.

PROBLEM

The snag is that the non-linear mixer stage produces both sum and difference frequency outputs, so that, for any given local oscillator frequency there will be two incoming signal frequencies at which reception would be possible.

These are known as signal and image frequencies, and it is essential that the selectivity of the tuned circuits preceding the mixer stage is adequate to reject these spurious second channel or image frequency

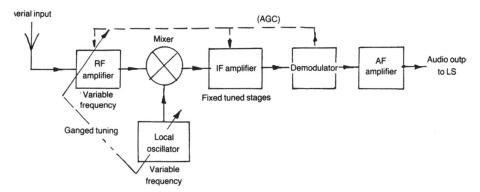

Fig. 2.21 *The superhet system.*

signals. This can be difficult if the IF frequency is too low in relation to the incoming signal frequency, since the image frequency will occur at twice the IF frequency removed from the signal, which may not be a very large proportion of the input signal frequency, bearing in mind the selectivity limitations of conventional RF tuned circuits.

In communications receivers and similar high-quality professional systems, the problem of image breakthrough is solved by the use of the double superhet system shown in Fig. 2.22. In this the incoming signal frequency is changed twice, firstly to a value which is sufficiently high to allow complete elimination of any spurious image frequency signals, and then, at a later stage, to a lower frequency at which bandwidth limitation and filtering can be done more easily.

A difficulty inherent in the superhet is that, because the mixer stage is, by definition, non-linear in its characteristics, it can cause intermodulation products between incoming signals to be generated, which will be amplified by the IF stages if they are at IF frequency, or will simply be combined with the incoming signal if they are large enough to drive the input of the mixer into overload.

This type of problem is lessened if the mixer device has a true square law characteristic. Junction FETs have the best performance in this respect.

A further problem is that the mixer stages tend to introduce a larger amount of noise into the signal chain than the other, more linear, gain stages, and this may well limit the ultimate sensitivity of the system. The noise figure of the mixer stage is partly a function of the kind of device or circuit configuration employed, and partly due to noise introduced by the relatively large amplitude local oscillator signal.

It is undoubtedly true to say that the quality and care in design of the mixer stage from a major determining factor in receiver performance, in respect of S/N ratio and freedom from spurious signals.

OTHER POSSIBILITIES

Because of the problems of the superhet system, in respect of mixer noise, and image channel interference, direct conversion systems have been proposed, in which the signal is demodulated by an electronic switch operated by a control waveform which is synchronous in frequency and phase with the input signal. These are known as homodyne or synchrodyne receivers, depending on whether the control waveform is derived from the carrier of the incoming signal or from a separate synchronous oscillator.

Since both of these systems result in an audio signal in which adjacent channel transmissions are reproduced at audio frequencies dependent on the difference of the incoming signal from the control frequency, they offer a means for the control of selectivity, with a truly flat-topped frequency response, by means of post-demodulator AF filtering. On the debit side, neither of these offer the sensitivity or the ease of operation of

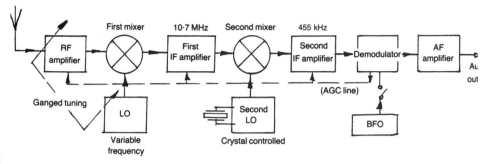

Fig. 2.22 *The double-superhet system, as used in a communication receiver.*

the conventional superhet, and are not used commercially to any significant extent.

Sensitivity

Many factors influence this characteristic. The most important of these are

- the S/N ratio, as depending on the design, small signals can get buried in mixer or other circuit noise;
- inadequate detectability, due to insufficient gain preceding the demodulator stage;
- intermodulation effects in which, due to poor RF selectivity or unsatisfactory mixer characteristics, the wanted signal is swamped by more powerful signals on adjacent channels.

The ultimate limitation on sensitivity will be imposed by aerial noise, due either to man-made interference, (RFI), or to the thermal radio emission background of free space. The only help in this case is a better aerial design.

As mentioned above, the frequency changer stage in a superhet is a weak link in the receiver chain, as it introduces a disproportionately high amount of noise, and is prone to intermodulation effects if the signals present exceed its optimum input signal levels.

The problem of mixer noise is a major one with equipment using thermionic valves, but semiconductor devices offer substantial improvements in this respect. For professional equipment, diode ring balanced modulator layouts, using hot carrier or Schottky diodes, of the kind shown in Fig. 2.23, are the preferred choice, since they combine excellent noise characteristics with the best possible overload margin. However, this is a complex system.

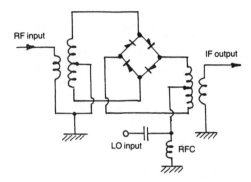

Fig. 2.23   *Diode ring double balanced mixer system.*

In domestic equipment, at frequencies up to about 100 MHz, junction FETs are the preferred devices, though they have inconveniently large inter-electrode capacitances for RF use. In the frequency range 100–500 MHz, dual-gate MOSFETs are preferable because their form of construction allows very good input – output screening, though their noise figure and other characteristics are somewhat less good than those of junction FETs.

Beyond 500 MHz, bipolar junction transistors are the only practical choice, though their use demands careful circuit design.

Integrated circuit balanced modulator systems have attracted some interest for high-quality designs, and have even been proposed as an alternative to ring diode modulators, though they have relatively poor overload characteristics. The various practical mixer systems are examined later, under 'Circuit design'.

In general, the best performance in a receiver, in respect of S/N ratio, sensitivity, and overload characteristics, requires a careful balance between the gain and selectivity of the various amplifying and mixing stages.

Stability

In a superhet system, in which there are a series of selective fixed-frequency amplifier stages, the major problems of frequency stability centre around the performance of the local (heterodyne) oscillator, which is combined with the incoming signal to give IF frequency. In relatively narrow bandwidth AM receivers, especially those operating in the short wave (3–30 MHz) region, a simple valve or transistor oscillator is unlikely to be adequately stable, unless very good quality components, and carefully balanced thermal compensation is employed.

Various techniques have been used to overcome this difficulty. For fixed frequency reception, oscillators based on individual quartz crystal or SAW resonators – which give an extremely stable output frequency, combined with high purity and low associated noise – are an excellent solution, though expensive if many reception frequencies are required. Alternatively, various ways of taking advantage of the excellent stability of the quartz crystal oscillator, while still allowing frequency variability, have been proposed, such as the phase-locked loop (PLL) frequency synthesiser, or the Barlow-Wadley loop systems.

QUARTZ CRYSTAL CONTROL

This operates by exciting a mechanical oscillation in a precisely dimensioned slab of mono-crystalline silica, either naturally occurring, as quartz, or, more commonly synthetically grown from an aqueous solution under conditions of very high temperature and pressure.

Since quartz exhibits piezo-electric characteristics, an applied alternating voltage at the correct frequency will cause the quartz slab to 'ring' in a manner analogous to that of a slab or rod of metal struck by a hammer. However, in the case of the crystal oscillator, electronic means can be used to sustain the oscillation.

The monotonic frequency characteristics of the quartz crystal resonator derive from its very high effective Q value. Typical apparent values of L_r, C_r and C_m (the resonant inductance and capacitance, and that due to the mounting), and series loss resistance R_1, for an 'X' cut crystal, are shown in the equivalent circuit of Fig. 2.24, for a crystal having a resonant frequency of 1 MHz, and an effective Q, as a series resonant circuit, of 300 000.

As in other materials, the physical dimensions of quartz crystals will change with temperature, but since its expansion is anisotropic (greater in some dimensions than others) it is possible to choose a particular section, or 'cut', through the crystal to minimise the effects of temperature on

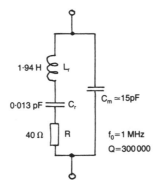

Fig. 2.24 *Equivalent electrical circuit of a quartz crystal resonator.*

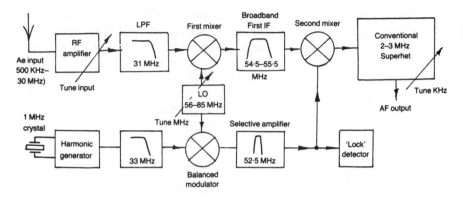

Fig. 2.25   *The Barlow-Wadley loop.*

resonant frequency. Such crystals are known as zero temperature coefficient (ZTC) or AT cut types.

Zero temperature coefficient crystals (in practice this will imply temperature coefficients of less than ±2 parts per million per degree centigrade over the temperature range 5−45° C) do not offer quite such high Q values as those optimised for this quality, so external temperature stabilisation circuits (known as crystal ovens) are sometimes used in critical applications. Both of these approaches may also be used, simultaneously, where very high stability is required.

Although, in principle, the quartz crystal is a fixed frequency device, it is possible to vary the resonant frequency by a small amount by altering the value of an externally connected parallel capacitor. This would be quite insufficient as a receiver tuning means, so other techniques have been evolved.

THE BARLOW-WADLEY LOOP

This circuit arrangement, also known as a drift cancelling oscillator, has been used for many years in relatively inexpensive amateur communications receivers, such as the Yaesu Musen FRG-7, and the layout adopted for such a 500 kHz–30 MHz receiver is shown in Fig. 2.25.

In this, the incoming signal, after appropriate RF amplification and pre-selection, is passed though a steep-cut, low-pass filter, which removes all signals above 31 MHz, to the first mixer stage. This has a conventional L-C type tuned oscillator whose operating frequency gives a first IF output in the range 54.5−55.5 MHz, which is fed to a second mixer stage.

The L-C oscillator output is also fed to a double-balanced modulator where it is combined with the output from 1 MHz quartz-crystal controlled harmonic generator, and this composite output is taken through a selective amplifier having a centre frequency of 52.5 MHz, and a signal output detector system.

Certain combinations of the local oscillator frequency and the harmonics of 1 MHz will be within the required frequency range, and will therefore pass through this amplifier. When such a condition exists, the output voltage operates an indicator to show that this signal is present. This output signal is then fed to the second mixer stage to generate a second IF frequency in the range 2–3 MHz, from which the desired signal is selected by a conventional superhet receiver from the 1 MHz slab of signals presented to it.

The frequency drift in the first, high-frequency, L-C local oscillator is thereby cancelled, since it will appear, simultaneously, and in the same sense, at both the first and second mixers.

FREQUENCY SYNTHESIZER TECHNIQUES

These are based on developments of the phase-locked loop (PLL) shown in Fig. 2.26. In this arrangement an input AC signal is compared in phase with the output from a voltage controlled oscillator (VCO). The output from the phase comparator will be the sum and difference frequencies of the input and VCO signals.

Where the difference frequency is low enough to pass the low-pass 'loop filter', the resultant control voltage applied to the VCO will tend to pull it into frequency synchronism, and phase quadrature, with the incoming signal – as long as the loop gain is high enough. In this condition, the loop is said to be 'locked'.

This circuit can be used to generate an AC signal in frequency synchronism with, but much larger in amplitude than, the incoming reference signal. It can also generate an oscillator control voltage which will accurately follow incoming variations in input signal frequency when the loop is in lock, and this provides an excellent method of extracting the modulation from an FM signal.

A further development of the basic PLL circuit is shown in Fig. 2.27. In this, a frequency divider is interposed between the VCO and the phase

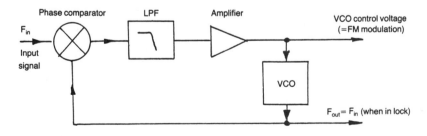

Fig. 2.26   *The phase-locked loop (PLL).*

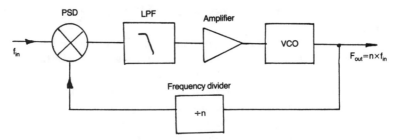

Fig. 2.27   *Phase-locked frequency multiplier.*

comparator, so that when the loop is in lock, the VCO output frequency will be a multiple of the incoming frequency. For example, if the divider has a factor n, then the VCO will have an output frequency equal to n (F_{in}).

In the PLL frequency synthesiser shown in Fig. 2.28, this process is taken one stage further, with a crystal controlled oscillator as the reference source, feeding the phase detector through a further frequency divider. If the two dividers have ratios of m and n, then when the loop is in lock, the output frequency will be $(n/m) \times F_{ref}$.

Provided that m and n are sufficiently large, the VCO output can be held to the chosen frequency, with crystal-controlled stability, and with any degree of precision required. Now that such frequency synthesiser circuitry is available in single IC form, this type of frequency control is beginning to appear in high quality FM tuners, as well as in communications receivers.

A minor operating snag with this type of system is that, because of the presence within the synthesiser IC of very many, relatively large amplitude, switching waveforms, covering a wide spectrum of frequencies, such receivers tend to be plagued by a multitude of minor tuning whistles, from which conventional single tuned oscillator systems are free. Very thorough screening of the synthesiser chip is necessary to keep these spurious signals down to an unobtrusive level.

Generally, in FM tuners, the relatively wide reception bandwidth makes oscillator frequency drift a less acute problem than in the case of AM receivers operating at comparable signal frequencies, although, since the distortion of the received signals will in most cases deteriorate if the set is not correctly tuned, or if it drifts off tune during use, there is still an incentive, in high-quality tuners, to employ quartz crystal stabilised oscillator systems.

A more serious problem, even in wide bandwidth FM tuners − where these do not employ PLL frequency control − is that the varicap diodes (semiconductor junction diodes in which the effective capacitance is an inverse function of the applied reverse voltage) often used to tune the RF

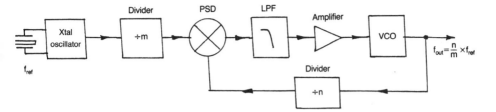

Fig. 2.28 *Phase-locked frequency synthesiser.*

and oscillator circuits, are quite strongly temperature dependant in their characteristics.

Varicap-tuned receivers must therefore employ thermally compensated DC voltage sources for their tuning controls if drift due to this cause is to be avoided.

AGC EFFECTS

A particular problem which can occur in any system in which automatic gain control (AGC) is used, is that the operation of the AGC circuit may cause a sympathetic drift in tuned frequency.

This arises because the AGC system operates by extracting a DC voltage which is directly related to the signal strength, at some point in the receiver where this voltage will be of adequate size. This voltage is then used to control the gain of preceding RF or IF amplifier stages so that the output of the amplifier remains substantially constant. This is usually done by applying the control voltage to one or other of the electrodes of the amplifying device so that its gain characteristics are modified.

Unfortunately, the application of a gain control voltage to any active device usually results in changes to the input, output, or internal feedback capacitances of the device, which can affect the resonant frequency of any tuned circuits attached to it. This effect can be minimised by care in the circuit design.

A different type of problem concerns the time-constants employed in the system. For effective response to rapid changes in the signal strength of the incoming signal, the integrating time constant of the system should be as short as practicable. However, if there is too little constraint on the speed of response of the AGC system, it may interpret a low-frequency modulation of the carrier, as for example in an organ pedal note, as a fluctuation in the received signal strength, and respond accordingly by increasing or decreasing the receiver gain to smooth this fluctuation out.

In general, a compromise is attempted between the speed of the AGC response, and the lowest anticipated modulation frequency which the receiver is expected to reproduce, usually set at 30 Hz in good quality receivers. In transistor portable radios, where the small LS units seldom reproduce tones below some 200−250 Hz, a much more rapid speed of response is usable without audible tonal penalties.

Sadly, some broadcasting authorities take advantage of the rapid AGC response typically found in transistor portables to employ a measure of companding (tonal range compression on transmission followed by expansion on reception to restore the original dynamic range).

As practised by the BBC on its Radio 1, Radio 2, Radio 4 and some local radio transmissions, this consists of a reduction in both carrier strength and modulation on all dynamic peaks. This reduces the amount of electricity consumed by the transmitter, whose average power output decreases on sound level peaks.

If, then, the AGC system in the radio restores the received carrier level to a constant value, the original modulation range will be recovered. This will only work well if the AGC 'attack' and 'decay' time constants used in the receiver are correctly related to those employed at the transmitter − and this is purely a matter of chance. The result, therefore, is an additional and unexpected source of degradation of the broadcast quality of these signals.

AUTOMATIC FREQUENCY CONTROL (AFC)

Because of the shape of the output voltage versus input frequency relationship, at the output of the demodulator of an FM receiver, shown in idealised form in Fig. 2.29, the possibility exists that this voltage, when averaged so that it is just a DC signal, with no carrier modulation, can be used to control the operating frequency of the oscillator, or other tuned circuits. So if the tuning of the receiver drifts away from the ideal mid-point (F_t in Fig. 2.29) an automatic correction can be applied to restore it to the desired value.

This technique is widely used, especially in the case of receivers where the tuning is performed by the application of a voltage to one or more varicap diodes, and which, in consequence, lends itself well to the superimposition of an additional AFC voltage. It does, however, have snags.

The first of these is that there is a close similarity in the action of an AFC voltage in an FM tuner to the action of an automatic gain control in an AM one. In both cases the control system sees the change − in carrier frequency in the case of an FM tuner − which is produced by the programme modulation, as being an error of the type which the control system is designed to correct. Too effective an AFC system can therefore lead to a worsening of bass response in the tuner, unless very heavy damping of the response is incorporated.

In the case of FM tuners, it is likely that the changes to be corrected will mainly be slow, and due only to thermal effects, provided that the receiver was correctly tuned in the first place, whereas in an AM radio the changes in received signal strength can often be quite rapid.

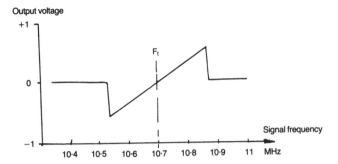

Fig. 2.29 *Voltage/frequency relationships in an ideal FM demodulator.*

The second problem is that the AFC system, attempting to sit the tuning point on the zero DC level, may not lead to the best results if the demodulator circuit is misaligned, and gives the kind of curve shown in Fig. 2.30. Here the user might place the tuning at the point F^1 by ear, where the AFC signal will always restore it to F_t.

This kind of demodulator misalignment is all too common in practice, and is one of the reasons why very linear demodulators (in which the choice of the correct point is less critical) are a worthwhile pursuit. Some more recent systems employ demodulator circuits which are comparatively insensitive to detuning. It should be remembered, also, that the output from the demodulator is likely to be used to operate the tuning meter, and encourage the user to tune to F_t.

In the case of this kind of tuning meter, the user can carry out a quick check on the alignment of the tuned circuits by noting the maximum meter deflection on either side of the incoming signal frequency. These readings should be symmetrical.

Predictability

The possession of a good, clear, tuning scale has always been a desirable virtue in any radio receiver, so that the user could know the point of tune and return without difficulty to the same place. However, with contemporary IC technology, the cost of integrated circuit frequency counter systems has become so low, relative to the other costs of the design, that almost all modern tuners now employ some kind of frequency meter display.

This operates in the manner shown in Fig. 2.31. In this, a quartz crystal controlled oscillator, operating perhaps at 32.768 kHz, (the standard frequency for 'quartz' watches, for which cheap crystals and simple frequency dividers are readily available — 32768 is 2^{15}) is used to generate a precise time interval, say one second.

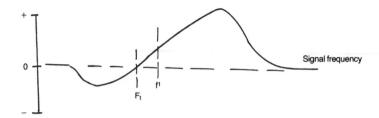

Fig. 2.30   *Effect of demodulator misalignment on AFC operation.*

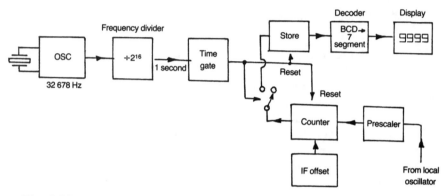

Fig. 2.31   *Frequency meter system.*

Meanwhile, the signal from the oscillator of the receiver, which will invariably employ a superhet layout, will be clocked over this period by an electronic counter, in which the IF frequency (that frequency by which the local oscillator differs from the incoming RF signal) will be added to, or subtracted from, the oscillator frequency — depending on whether the oscillator frequency is higher or lower than that of the signal — so that the counter effectively registers the incoming signal frequency.

The electronic switch circuitry operated by the time interval generator is then used alternately to reset the frequency counter to zero, and to transfer its final count to a digital frequency store. A liquid crystal (LCD) or light-emitting diode (LED) display can then be used to register the signal frequency in numerical form.

As a matter of convenience, to avoid display flicker the counter output will be held in a store or latch circuit during the counting period, and the contents of the store only updated each time the count is completed. Since frequency dither is only likely to affect the last number in the display, in circumstances where there is an uncertainty of ± 1 digit, some systems actually count to a larger number than is displayed, or up-date the last — least significant — digit less frequently than the others.

Additional information can also be shown on such an LED/LCD display, such as the band setting of the receiver, whether it is on AM or FM, whether AFC or stereo decoding has been selected, or the time. In the light of the continual search for marketable improvements, it seems probable that such displays will soon also include a precise time signal derived from one or other of the time code transmitters.

Clarity

One of the major advantages of FM radio is its ability to reject external radio frequency noise – such as that due to thunderstorms, which is very prominent on LF radio, or motor car ignition noise, which is a nuisance on all HF and VHF transmissions, which includes Band 2.

This ability to reject such noise depends on the fact that most impulse noise is primarily amplitude modulated, so that, while it will extend over a very wide frequency range, its instantaneous frequency distribution is constant. The design of the FM receiver is chosen, deliberately, so that the demodulator system is insensitive to AM inputs, and this AM rejection quality is enhanced by the use of amplitude limiting stages prior to the demodulator.

A valuable 'figure of merit' in an FM receiver is its AM rejection ratio, and a well-designed receiver should offer at least 60 dB (1000:1).

Because FM receivers are invariably designed so that there is an excess of gain in the RF and IF stages, so that every signal received which is above the detection threshold will be presented to the demodulator as an amplitude limited HF square wave, it is very seldom that such receivers – other than very exotic and high-cost designs – will employ AGC, and only then in the pre-mixer RF stage(s).

A further benefit in well-designed FM receivers is that intruding AM radio signals on the same band will be ignored, as will less strong FM broadcasts on the same channel frequency. This latter quality is referred to as the capture ratio, and is expressed as the difference in the (voltage) signal strength, in dB, between a more powerful and a less powerful FM transmission, on the same frequency allocation, which is necessary for the weaker transmission to be ignored.

The capture ratio of an FM receiver – a desirable feature which does not exist in AM reception – depends on the linearity, both in phase and amplitude, of the RF and IF stages prior to limiting, and also, to a great extent, on the type of demodulator employed. The best current designs offer 1 dB discrimination. Less good designs may only offer 3–4 dB. Two decibels is regarded as an acceptable figure for high-quality systems, and would imply that the designers had exercised suitable care.

It should be noted in this context that designs in which very high IF gain is employed to improve receiver sensitivity may, by causing overload in these stages, and consequent cross-modulation if they do not 'clip' cleanly, lead to a degradation in capture ratio.

Needless to say, those IF stages which are designed to limit the amplitude of the incoming signal should be designed so that they do not generate inter-modulation products. However, these invariably take the form of highly developed IC designs, based on circuit layouts of the form shown in Fig. 2.32, using a sequence of symmetrically driven, emitter-coupled 'long-tailed pairs', chosen to operate under conditions where clean and balanced signal clipping will occur.

FM tuner designers very seldom feel inspired to attempt to better their performance by the use of alternative layouts.

In AM receivers, the best that can be done to limit impulse type interference is to use some form of impulse amplitude limiter, which operates either on the detection of the peak amplitude of the incoming signal or on the rate-of-change of that amplitude, such qualities will be higher on impulse noise than on normal programme content. An excellent AM noise limiter circuit from Philips, which incorporates an electronic delay line to allow the noise limiter to operate before the noise pulse arrives, is shown in Fig. 2.33.

The rejection of intruding signals in AM depends entirely on the receiver selectivity, its freedom from intermodulation defects, and − in respect of its ability to avoid internally generated hum, noise, and mush − on the quality of the circuit design and components used, and the wisdom of the choice of distribution of the gain and selectivity within the RF and IF stages.

Linearity, FM systems

In view of the continuing search for improved audio amplifier quality in both quantitative and subjective terms, it is not surprising that there has been a comparable effort to obtain a high performance in FM receiver systems.

This endeavour is maintained by competitive rivalry and commercial pressures, and has resulted in many cases in the development of FM tuners with a quality which exceeds that of the incoming signal presented to them by the broadcasting authorities. Broadcasters' activities are not the subject of competitive pressures, and their standards are determined by the reluctance of governments to spend money on luxuries, and by the cynicism of their engineers.

These constraints have resulted in a continuing erosion of the quality of AM radio broadcasts, though there still remain some honourable exceptions. This has, sadly, often led to the AM sections of FM tuners being designed as low-cost functional appendages having a very poor audio performance, even when the quality of the FM section is beyond reproach.

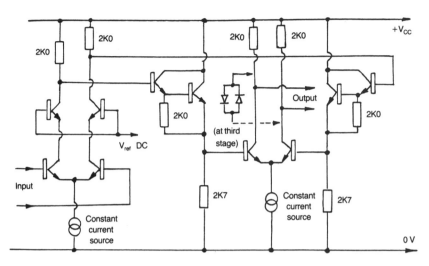

Fig. 2.32 *Cascode input stage, and the first (of two) symmetrical gain stages in a modern FM IF gain block IC (RCA CA3189E).*

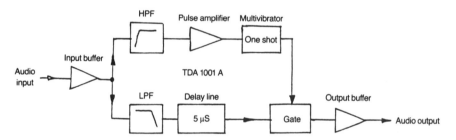

Fig. 2.33 *Philips' impulse noise limiting circuit.*

A number of factors influence the quality of the demodulated signal, beginning with the stability and phase linearity of the RF, IF and AF gain stages. In the case of AM radios, incipient instability in the RF or IF stages will lead to substantial amplitude distortion effects, with consequent proneness to intermodulation defects. (Any non-linearity in an amplifier will lead to a muddling of the signals presented to it.) In the case of those FM radios based on a phase-sensitive demodulator, any RF or IF instability will lead to substantial phase distortions as the signal passes through the frequency of incipient oscillation.

Much care is therefore needed in the design of such stages to ensure their stable operation. If junction FETs are employed rather than dual-gate MOSFETs, some technique, such as that shown in Fig. 2.34, must be used to neutralise the residual drain-gate capacitance. The circuit of Fig. 2.34 utilises a small inductance in the source lead, which could

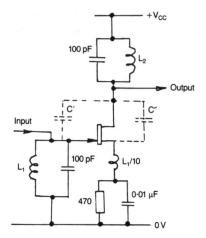

Fig. 2.34 *Feedback neutralisation system for junction FET RF amplifier.*

be simply a lengthy track on the printed circuit board, to ensure that the unwanted feedback signal due to the gate-drain capacitance (C'), is cancelled by a signal, effectively in phase opposition, due to the drain-source capacitance (C'').

However, assuming competent RF/IF stage design, the dominant factor in recovered signal quality is that of the demodulator design. Representative demodulator systems, for FM and for AM, are shown below.

SLOPE DETECTION

This circuit, shown in Fig. 2.35(a), is the earliest and crudest method of detecting or demodulating an FM signal. The receiver is simply tuned to one side or the other of the tuned circuit resonant frequency (F_0), as illustrated in Fig. 2.35(b). Variations in the incoming frequency will then produce changes in the output voltage of the receiver, which can be treated as a simple AM signal. This offers no AM rejection ability, and is very non-linear in its audio response, due to the shape of the resonance curve.

THE ROUND–TRAVIS DETECTOR

This arrangement, shown in Fig. 2.36(a), employes a pair of tuned circuits with associated diode rectifiers ('detectors') which are tuned, respectively, above and below the incoming signal frequency, giving a balanced slope-detector characteristic, as seen in Fig. 2.36(b). This is more linear than the simple slope-detector circuit, but still gives no worthwhile AM rejection.

THE FOSTER–SEELEY OR PHASE DETECTOR

This circuit, of which one form is shown in Fig. 2.37(a), was evolved to

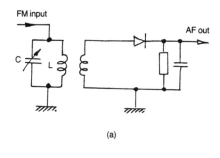

(a)

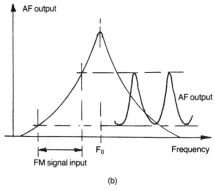

(b)

Fig. 2.35 *FM slope detector.*

provide an improved measure of AM rejection, by making its output dependent, at least in part, on the changes in phase induced in a tuned circuit by variations in the frequency of the incoming signal.

In this arrangement the tuned circuit, L_3C_1, provides a balanced drive to a matched pair of diode rectifiers (D_1, D_2), arranged in opposition so that any normal AM effects will cancel out. A subsidiary coil, L_2, is then arranged to feed the centre tap of L_3, so that the induced signal in L_2, which will vary in phase with frequency, will either reinforce or lessen the voltages induced in each half of L_3, by effectively disturbing the position of the electrical centre tap.

This gives the sort of response curve shown in Fig. 2.37(b) which has better linearity than its predecessors.

THE RATIO DETECTOR

This circuit, of which one form is shown in Fig. 2.38(a) is similar to the Foster–Seeley arrangement, except that the diode rectifiers are connected so that they produce an output voltage which is opposed, and balanced across the load. The output response, shown in Fig. 2.38(b), is very similar to that of the Foster–Seeley circuit, but it has a greatly improved AM rejection.

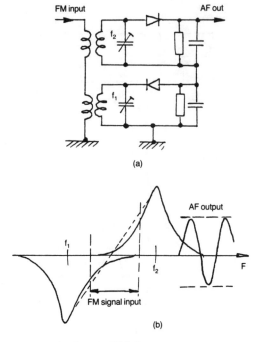

Fig. 2.36   *The Round–Travis FM detector.*

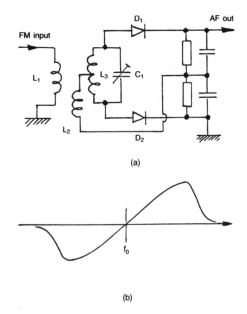

Fig. 2.37   *The Foster–Seeley or phase detector.*

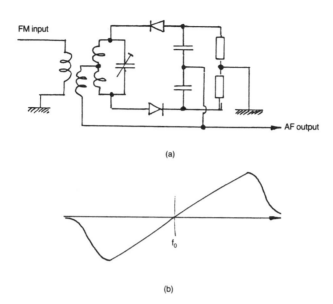

FM input

AF output

(a)

f_0

(b)

Fig. 2.38 *The ratio detector.*

The ratio detector was for many years the basic demodulator circuit for FM receivers, and offers a very good internal noise figure. This is superior to that of the contemporary IC-based phase-coincidence system, which has entirely superseded it, because of the much greater demodulation linearity offered by this latter system.

THE PHASE COINCIDENCE DEMODULATOR (PCD)

This method employs a circuit layout of the general form shown in Fig. 2.39, in which a group of identical bipolar transistors is interconnected so that the current from Q_1, a simple constant-current source, will be routed either through Q_2 or Q_3, depending on the relative potential of the signal.

From Q_2/Q_3, the current flow will be directed either through Q_4/Q_7 or through Q_5/Q_6 and recombined at the load resistors R_1 and R_2.

It will be seen, from inspection, that if the transistors are well matched, the potential drop across R_1 and R_2 will be that due to half the output current of Q_1, regardless of the relative potentials applied to Q_2 or Q_3 or to Q_4/Q_7 or Q_5/Q_6, so long as these potentials are not simultaneously applied.

If synchronous HF signals are applied to all four input ports (a−d), the output across R_1 or R_2, (output ports e and f) will only be identical if inputs c and d are at phase quadrature to those at a and b.

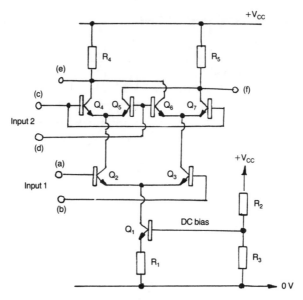

Fig. 2.39 *Gate-coincidence transistor array.*

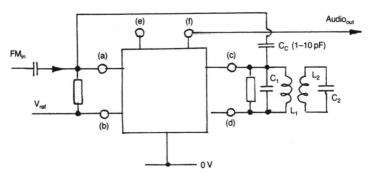

Fig. 2.40 *Phase coincidence demodulator circuit.*

Considering the circuit layout of Fig. 2.40, if ports b and d are taken to some appropriate DC reference potential, and an amplitude limited FM signal is applied to point a with respect to b, and some identical frequency signal, at phase quadrature at F_o, is applied to c with respect to d, then there will be an output voltage at point e with respect to point f, if the input frequency is varied in respect of F_o.

The linearity of this V/F relationship is critically dependent on the short-term frequency/phase stability of the potential developed across the quadrature circuit (L_1C_1), and this depends on the Q value of the tuned circuit. The snag here is that too high a value of Q will limit the usable FM bandwidth. A normal improvement which is employed is the addition of a further tuned circuit (L_2C_2) to give a bandpass coupling characteristic, and further elaborations of this type are also used to improve the linearity, and lessen the harmonic distortion introduced by this type of demodulator.

The performance of this type of demodulator is improved if both the signal drive (to ports a and b) and the feed to the quadrature circuit (ports c and d) are symmetrical, and this is done in some high quality systems.

THE PHASE-LOCKED LOOP (PLL) DEMODULATOR

This employs a system of the type shown above in Fig. 2.26. If the voltage controlled oscillator (VCO) used in the loop has a linear input voltage versus output frequency characteristic, then when this is in frequency lock with the incoming signal the control voltage applied to the VCO will be an accurate replica of the frequency excursions of the input signal, within the limits imposed by the low-pass loop filter.

This arrangement has a great advantage over all of the preceding demodulator systems in that it is sensitive only to the instantaneous input frequency, and not to the input signal phase. This allows a very low demodulator THD, unaffected by the phase-linearity of preceding RF and IF tuned circuits or SAW filters, and greatly reduces the cost, for a given performance standard, of the FM receiver system.

Such a circuit also has a very high figure for AM rejection and capture ratio, even when off-tune, provided that the VCO is still in lock.

Photographs of the output signal distortion, and the demodulator output voltage versus input frequency curves, taken from a frequency modulated oscillator display, are shown in Figs 2.41–2.43, for actual commercial receivers employing ratio detector and phase coincidence demodulator systems, together with the comparable performance of an early experimental PLL receiver due to the author.

All of these units had been in use for some time, and showed the effects of the misalignment which could be expected to occur with the passage of time. When new, both the conventional FM tuners would have probably given a better performance than at the time of the subsequent test. However, the superiority of the PLL system is evident.

Commercial manufacturers of domestic style equipment have been slow to exploit the qualities of the PLL, perhaps deterred by the unpleasant audio signal generated when the tuner is on the edge of lock. It is not, though, a difficult matter to incorporate a muting circuit which cuts off the signal when the receiver is off tune, and an experimental system of this kind has been in use for many years.

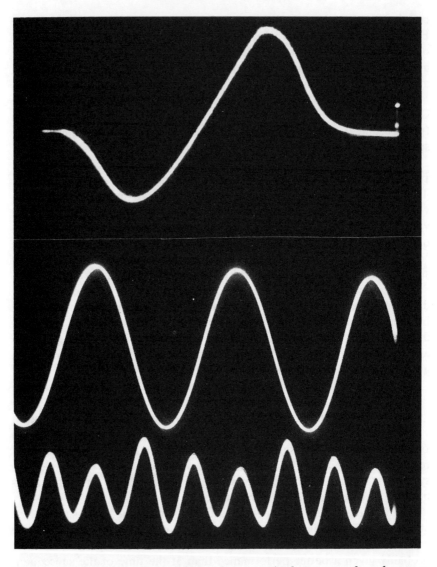

Fig. 2.41 *Practical demodulator frequency/voltage transfer characteristics, recovered signal, and overall receiver distortion waveform, (THD = 1.5%, mainly third harmonic), for ratio detector.*

PULSE COUNTING SYSTEMS

In the early years of FM transmissions, when there was great interest in the exploitation of the very high quality signals then available, pulse counting systems were commonly employed as the basis for high fidelity

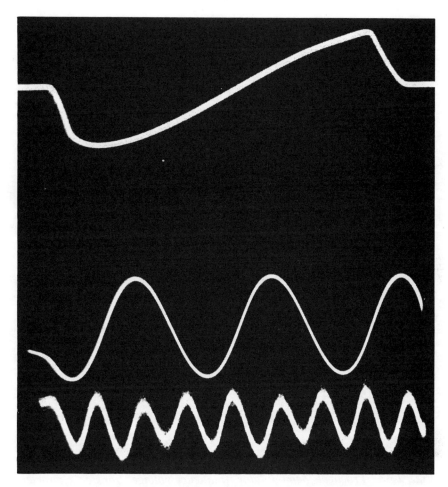

Fig. 2.42 *Practical demodulator frequency/voltage transfer character-istics, recovered signal, and overall receiver distortion waveform, (THD = 0.6%, mainly third harmonic), for phase coincidence FM demodulator.*

amateur designs. Typical circuit arrangements employed were of the form shown in Fig. 2.44. After suitable RF amplification, the incoming signal would be mixed with a crystal controlled local oscillator signal, in a conventional superhet layout, to give a relatively low frequency IF in, say, the range 30–180 kHz, which could be amplified by a conventional broad bandwidth HF amplifier.

The output signal would then be analysed by a linear rate meter circuit, to give a DC output level which was dependent on the instantaneous input signal frequency, yielding a low distortion recovered AF output.

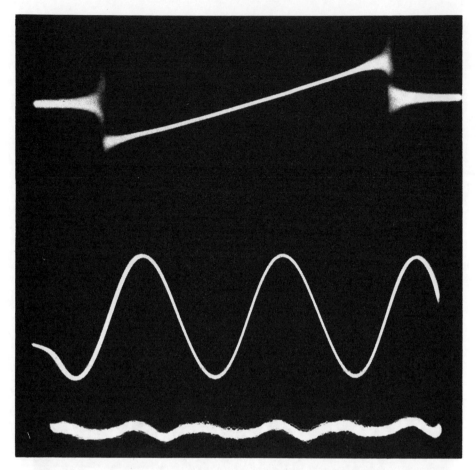

Fig. 2.43 *Practical demodulator frequency/voltage transfer character-*
istics, recovered signal, and overall receiver distortion waveform, (THD =
0.15%, mainly second harmonic), for phase-locked loop FM demodulator.

Such systems shared the quality of the PLL demodulator that the
output signal linearity was largely unaffected by the frequency/phase
linearity of the preceding RF/IF circuitry. Unfortunately, they did not
lend themselves well to the demodulation of stereo signals, and the
method fell into disuse.

However, this technique has been resurrected by Pioneer, in its F-90/
F-99 receivers, in a manner which exploits the capabilities of modern
digital ICs.

The circuit layout employed by Pioneer is shown in schematic form in
Fig. 2.45. In this the incoming 10.7 MHz signal is frequency doubled,
to double the modulation width, and mixed down to 1.26 MHz with a

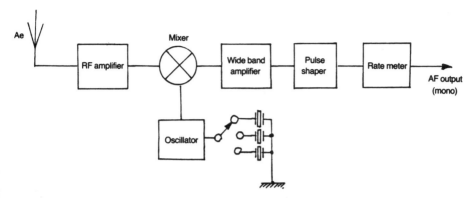

Fig. 2.44 *Pulse counting FM tuner.*

stable crystal controlled oscillator. After filtering, the signal is cleaned up and converted into a series of short duration pulses, having constant width and amplitude, of which the average value is the desired composite (L+R) audio signal.

A conventional PLL arrangement of the type shown in Fig. 2.27, is then used to reconstruct the 38 kHz sub-carrier signal, from which the LH and RH stereo outputs can be derived by adding the (L+R) + (L−R) and (L+R) + (R−L) components.

Practical FM receiver performance data

The claimed performance from the Pioneer F-90 receiver, using this pulse counting system, is that the THD is better than 0.01% (mono) and 0.02% (stereo), with a capture ratio of better than 1 dB, and a stereo channel separation greater than 60 dB. These figures are substantially better than any obtainable with more conventional demodulator systems.

In general, the expected THD performance for a good modern FM receiver is 0.1% (mono) and 0.2−0.3% (stereo), with capture ratios in the range 1 dB (excellent) to 2.5 dB (adequate), and stereo separations in the range 30−50 dB. These figures largely depend on the type of demodulator employed, and the quality of the RF and IF circuit components and alignment.

Typical AF bandwidths will be in the range 20−40 Hz to 14.5 kHz (−3 dB points). The LF frequency response depends on the demodulator type, with PLL and pulse counting systems allowing better LF extension. The upper frequency limit is set by the requirements of the Zenith-GE encoding system, rather than by the receiver in use.

The ultimate signal to noise ratio of a good receiver could well be of

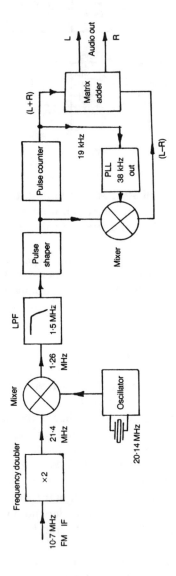

Fig. 2.45 *The Pioneer pulse counting FM tuner system.*

the order of 70 dB for a mono signal, but normal reception conditions will reduce this figure somewhat, to 60 dB or so.

Input (aerial) signal strengths greatly influence the final noise figure, which worsens at low signal levels, so that input receiver sensitivities of the order of 2.5 μV and 25 μV might be expected for a 50 dB final S/N figure, on mono and stereo signals respectively, from a first quality receiver. Values of 25 μV/100 μV might be expected from less good designs.

A high (aerial) input sensitivity and a good intrinsic receiver S/N ratio is of value if the receiver is to be used with poor or badly sited aerial systems, even though the difference between receivers of widely different performance in this respect may not be noticeable with a better aerial installation.

Linearity, AM systems

The performance of an AM radio is invariably less good than that possible from a comparable quality FM receiver. The reasons for this are partly to do with the greater difficulty in obtaining a low demodulator distortion level and a wide AF bandwidth, with a sensibly flat frequency response, coupled with good signal to noise and selectivity figures − a difficulty which is inherent in the AM system − and partly to do with the quality of the radio signal, which is often poor, both as received and as transmitted.

There are, however, differences between the various types of AM demodulator, and some of these have significant performance advantages over those normally used in practice. These are examined below.

THE DIODE 'ENVELOPE' DEMODULATOR

This is a direct descendant of the old crystal detector of the 'crystal and cat's-whisker' era, and is shown in Fig. 2.46. In this the peak voltage level occurring across the tuned circuit L_2C_1 is passed through the rectifier

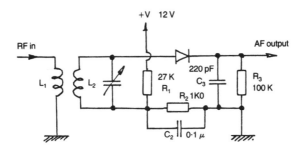

Fig. 2.46 *Simple forward-biased diode AM detector.*

diode, D_1, and stored in the output filter circuit C_3R_3. Since the rectifying diode will require some forward voltage to make it conduct (about 0.15 V in the case of germanium, and 0.55 V in the case of silicon types) it is good practice to apply a forward bias voltage, through $R_1R_2C_2$, to bring the diode to the threshold of forward conduction.

Because it is essential that the RF component is removed from the output signal, there is a minimum practical value for C_3. However, since this holds the peak audio modulation voltage until it can decay through R_3, a measure of waveform distortion is inevitable, especially at higher end of the audio range and at lower RF input signal frequencies.

Typical performance figures for such demodulators are 1–2% at 1 kHz.

'GRID-LEAK' DEMODULATION
This was a common system used in the days of amateur radio receivers, and shown in its thermionic valve form in Fig. 2.47(a). A more modern version of this arrangement, using a junction FET, is shown in Fig. 2.47(b). Both these circuits operate by allowing the RF signal appearing across the tuned circuit L_2C_1 to develop a reverse bias across R_1, which reduces the anode or drain currents.

The THD performance of such circuits is similar to that of the forward-biased diode demodulator, but they have a lower damping effect on the preceding tuned circuits.

ANODE-BEND OR INFINITE IMPEDANCE DEMODULATOR SYSTEMS
These are shown in their junction FET versions, in Fig 2.48(a) and 2.48(b). They are similar in their action and only differ in the position of the output load. They operate by taking advantage of the inevitable curvature of the I_d/V_g curve for a valve or FET to provide a distorted version of the incoming radio signal, as shown in Fig. 2.49, which, when averaged by some integrating circuit, gives an audio signal equivalent to the modulation.

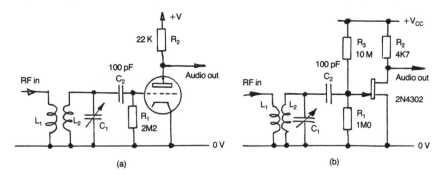

Fig. 2.47   *Valve and FET grid-leak AM detectors.*

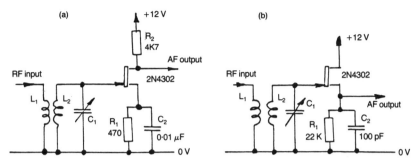

Fig. 2.48 *FET based anode-bend and infinite impedance detectors.*

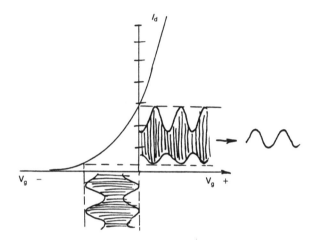

Fig. 2.49 *Method of operation of anode-bend or infinite impedance detectors.*

Since the circuit distorts the incoming RF waveform, it is unavoidable that there will be a related distortion in the recovered AF signal too. Typical THD figures for such a demodulator system will lie in the range 0.5–1.5%, depending on signal level. Most IC AM radio systems will employ either diode envelope detectors or variations of the anode bend arrangement based on semiconductor devices.

The performance of all these systems can be improved by the addition of an unmodulated carrier, either at the same frequency, or at one substantially different from it, to improve the carrier to modulation ratio. This is not a technique which is used for equipment destined for the domestic market, in spite of the quality improvement possible.

SYNCHRONOUS DEMODULATION

This method, illustrated schematically in Fig. 2.50(a), operates by

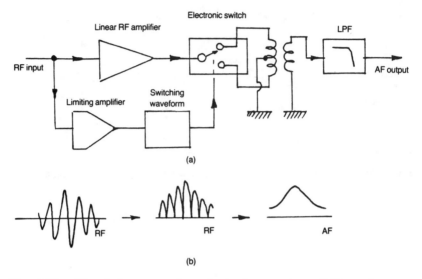

(b)

Fig. 2.50 *Circuit layout and method of operation of Homodyne type synchronous demodulator.*

synchronously inverting the phase of alternate halves of the RF signal, so that the two halves of the modulated carrier can be added together to give a unidirectional output. Then, when the RF component is removed by filtering, only the audio modulation waveform remains, as shown in Fig. 2.50(b).

This technique is widely used in professional communication systems, and is capable of THD values well below 0.1%.

CIRCUIT DESIGN

Although there is a continuing interest in circuit and performance improvements made possible by new circuit and device technology, there is also a tendency towards a degree of uniformity in circuit design, as certain approaches establish themselves as the best, or the most cost-effective.

This is particularly true in the RF and IF stages of modern FM tuners, and an illustration of the type of design approach employed is given in Fig. 2.51. The only feature in this not covered by the previous discussion is the general use of dual (back-to-back) Varicap diodes. These are preferred to single diodes as the tuning method in high-quality systems because they avoid the distortion of the RF waveform brought about by the signal voltage modulating the diode capacitance.

Variable air-spaced ganged capacitors would be better still, but these

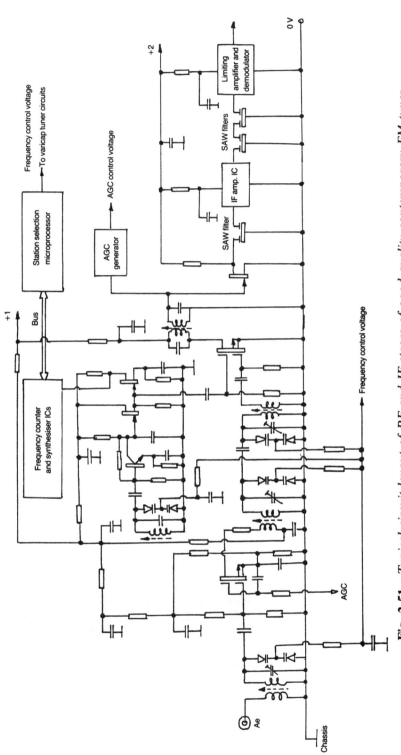

Fig. 2.51 *Typical circuit layout of RF and IF stages of good quality contemporary FM tuner.*

are bulky and expensive, and do not lend themselves to frequency synthesiser layouts.

NEW DEVELOPMENTS

In such a fast changing market it is difficult to single out recent or
projected design features for special mention, but certain trends are
apparent. These mainly concern the use of microprocessor technology to
memorise and store user selections of channel frequencies, as in the Quad
FM4 (Fig. 2.52), and the use of 'sliding stereo separation' systems to
lessen the $L-R$ channel separation, with its associated 'stereo hiss', when
the incoming signal strength falls below the optimum value.

Some synthesiser tuners offer normal spin-wheel tuning, as in the
Harmon–Kardon TU915, by coupling the tuning knob shaft to the synthesiser IC by the use of an optical shaft encoder.

Low-noise gallium arsenide dual-gate Mosfets have made an appearance
in the Hitachi FT5500 receiver, and these devices are likely to be more
widely adopted in such systems.

Clearly, commercial pressures will encourage manufacturers to develop
more complex large-scale integration (LSI) integrated circuits, so that
more of the receiver circuitry may be held on a single chip. This undoubtedly saves manufacturing costs, but the results are not always of benefit to
the user, as evidenced by the current performance of single IC AM radio
sections.

An increasing number of the better FM tuners are now offering a
choice of IF bandwidths, to permit user optimisation of selectivity, sensitivity or stereo separation. Variable receiver bandwidth would be a valuable feature on the MF bands, and may be offered if there is any serious
attempt to improve the quality of this type of receiver.

Fig. 2.52 *The Quad FM tuner.*

APPENDIX 2.1: BROADCAST SIGNAL CHARACTERISTICS

(Data by courtesy of the BBC)

Audio bandwidths

MF:
- 40–5800 Hz ±3 dB, with very sharp cut off. (> 24 dB/octave beyond 6 kHz.)
- 50–5000 Hz ±1 dB.

VHF:
- 30–15,000 Hz ±0.5 dB, with very sharp cut off beyond this frequency.

Distortion

MF:
- < 3% THD at 75% modulation.
- < 4% THD at 100% modulation.

VHF:
- < 0.5% THD.

Stereo crosstalk

VHF: > 46 dB. (0.5%).

Modulation depth

MF: Up to 100% over the range 100–5000 Hz.
VHF: Peak deviation level corresponds to ±60.75 kHz deviation. (The total deviation, including pilot tone, is ±75 kHz.)

S/N ratio

MF: > 54 dB below 100% modulation.
VHF: Up to 64 dB, using CCIR/468 weighting, with reference to peak programme modulation level.

APPENDIX 2.2: RADIO DATA SYSTEM (RDS)

(Data by courtesy of the BBC)

It was proposed to begin the introduction of this system in the Autumn of 1987, and when in use on receivers adapted to receive this signal the reception of station, programme and other data is possible.

This data will include programme identification, to allow the receiver to automatically locate and select the best available signal carrying the chosen programme, and to display in alphanumeric form a suitable legend, and to permit display of clock time and date.

Anticipated future developments of this system include a facility for automatic selection of programme type (speech/light music/serious music/ news), and for the visual display of information, such as traffic conditions, phone-in numbers, news flashes, programme titles or contents, and data print out via a computer link.

CHAPTER 3

Preamplifiers and input signals

REQUIREMENTS

Most high-quality audio systems are required to operate from a variety of signal inputs, including radio tuners, cassette or reel-to-reel tape recorders, compact disc players and more traditional record player systems. It is unlikely at the present time that there will be much agreement between the suppliers of these ancillary units on the standards of output impedance or signal voltage which their equipment should offer.

Except where a manufacturer has assembled a group of such units, for which the interconnections are custom designed and there is in-house agreement on signal and impedance levels – and, sadly, such ready-made groupings of units seldom offer the highest overall sound quality available at any given time – both the designer and the user of the power amplifier are confronted with the need to ensure that his system is capable of working satisfactorily from all of these likely inputs.

For this reason, it is conventional practice to interpose a versatile pre-amplifier unit between the power amplifier and the external signal sources, to perform the input signal switching and signal level adjustment functions.

This pre-amplifier either forms an integral part of the main power amplifier unit or, as is more common with the higher quality units, is a free-standing, separately powered, unit.

SIGNAL VOLTAGE AND IMPEDANCE LEVELS

Many different conventions exist for the output impedances and signal levels given by ancillary units. For tuners and cassette recorders, the output is either be that of the German DIN (Deutsches Industrie Normal) standard, in which the unit is designed as a current source which will give an output voltage of 1 mV for each 1000 ohms of load impedance, such that a unit with a 100 K input impedance would see an input signal voltage of 100 mV, or the line output standard, designed to drive a load of 600 ohms or greater, at a mean signal level of 0.775 V RMS, often referred to in tape recorder terminology as OVU.

119

Generally, but not invariably, units having DIN type interconnections, of the styles shown in Fig. 3.1, will conform to the DIN signal and impedance level convention, while those having 'phono' plug/socket outputs, of the form shown in Fig. 3.2 will not. In this case, the permissible minimum load impedance will be within the range 600 ohms to 10000 ohms, and the mean output signal level will commonly be within the range 0.25–1 V RMS.

An exception to this exists in respect of compact disc players, where the output level is most commonly 2 V RMS.

GRAMOPHONE PICK-UP INPUTS

Three broad categories of pick-up cartridge exist: the ceramic, the moving magnet or variable reluctance, and the moving coil. Each of these has different output characteristics and load requirements.

Ceramic piezo-electric cartridges

These units operate by causing the movement of the stylus due to the groove modulation to flex a resiliently mounted strip of piezo-electric ceramic, which then causes an electrical voltage to be developed across metallic contacts bonded to the surface of the strip. They are commonly

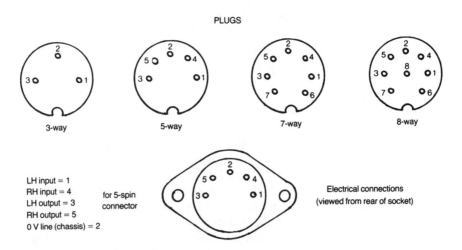

Fig. 3.1 *Common DIN connector configurations.*

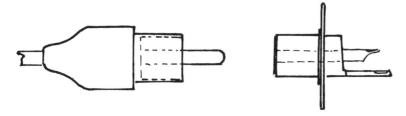

Fig. 3.2 *The phono connector.*

found only on low-cost units, and have a relatively high output signal level, in the range 100–200 mV at 1 kHz.

Generally the electromechanical characteristics of these cartridges are tailored so that they give a fairly flat frequency response, though with some unavoidable loss of HF response beyond 2 kHz, when fed into a pre-amplifier input load of 47 000 ohms.

Neither the HF response nor the tracking characteristics of ceramic cartridges are particularly good, though circuitry has been designed with the specific aim of optimising the performance obtainable from these units (see Linsley Hood, J., *Wireless World*, July 1969). However, in recent years, the continuing development of pick-up cartridges has resulted in a substantial fall in the price of the less exotic moving magnet or variable reluctance types, so that it no longer makes economic sense to use ceramic cartridges, except where their low cost and robust nature are of importance.

Moving magnet and variable reluctance cartridges

These are substantially identical in their performance characteristics, and are designed to operate into a 47 K load impedance, in parallel with some 200–500 pF of anticipated lead capacitance. Since it is probable that the actual capacitance of the connecting leads will only be of the order of 50–100 pF, some additional input capacitance, connected across the phono input socket, is customary. This also will help reduce the probability of unwanted radio signal breakthrough.

PU cartridges of this type will give an output voltage which increases with frequency in the manner shown in Fig. 3.3(a), following the velocity characteristics to which LP records are produced, in conformity with the RIAA recording standards. The pre-amplifier will then be required to have a gain/frequency characteristic of the form shown in Fig. 3.3(b), with the de-emphasis time constants of 3180, 318 and 75 microseconds, as indicated in the figure.

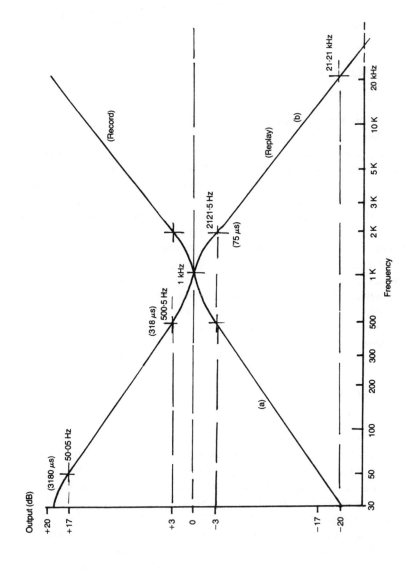

Fig. 3.3 *The RIAA record/replay characteristics used for 33/45 rpm vinyl discs.*

The output levels produced by such pick-up cartridges will be of the order of 0.8—2 mV/cm/s of groove modulation velocity, giving typical mean outputs in the range of 3—10 mV at 1 kHz.

Moving coil pick-up cartridges

These low-impedance, low-output PU cartridges have been manufactured and used without particular comment for very many years. They have come into considerable prominence in the past decade, because of their superior transient characteristics and dynamic range, as the choice of those audiophiles who seek the ultimate in sound quality, even though their tracking characteristics are often less good than is normal for MM and variable reluctance types.

Typical signal output levels from these cartridges will be in the range 0.02—0.2 mV/cm/s, into a 50—75 ohm load impedance. Normally a very low-noise head amplifier circuit will be required to increase this signal voltage to a level acceptable at the input of the RIAA equalisation circuitry, though some of the high output types will be capable of operating directly into the high-level RIAA input. Such cartridges will generally be designed to operate with a 47 K load impedance.

INPUT CIRCUITRY

Most of the inputs to the pre-amplifier will merely require appropriate amplification and impedance transformation to match the signal and impedance levels of the source to those required at the input of the power amplifier. However, the necessary equalisation of the input frequency response from a moving magnet, moving coil or variable reluctance pick-up cartridge, when replaying an RIAA pre-emphasised vinyl disc, requires special frequency shaping networks.

Various circuit layouts have been employed in the preamplifier to generate the required 'RIAA' replay curve for velocity sensitive pick-up transducers, and these are shown in Fig. 3.4. Of these circuits, the two simplest are the 'passive' equalisation networks shown in (a) and (b), though for accuracy in frequency response they require that the source impedance is very low, and that the load impedance is very high in relation to R_1.

The required component values for these networks have been derived by Livy (Livy, W.H., *Wireless World*, Jan. 1957, p. 29) in terms of RC time constants, and set out in a more easily applicable form by Baxandall (P. J. Baxandall, *Radio, TV and Audio Reference Book*, S.W. Amos [ed.], Newnes-Butterworth Ltd., Ch. 14), in his analysis of the various possible equalisation circuit arrangements.

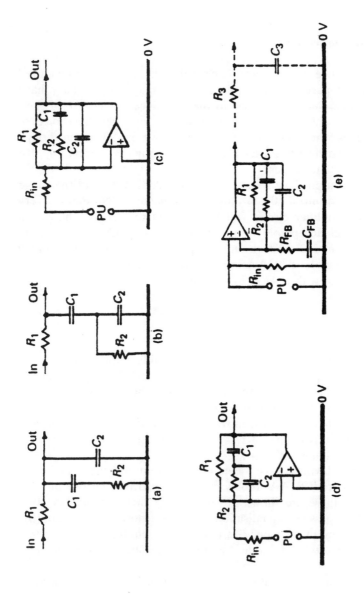

Fig. 3.4 *Circuit layouts which will generate the type of frequency response required for RIAA input equalization.*

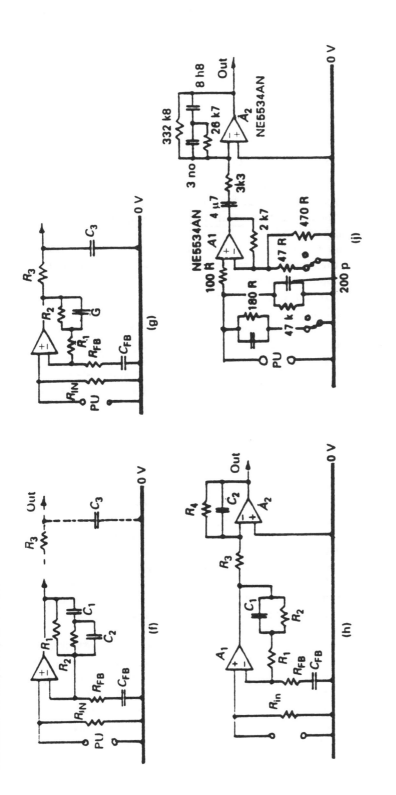

Fig. 3.4 *cont.*

From the equations quoted, the component values required for use in the circuits of Figs 3.4(a) and (c), would be:

$$R_1/R_2 = 6.818 \qquad C_1.R_1 = 2187 \text{ } \mu\text{s} \qquad \text{and } C_2.R_2 = 109 \text{ } \mu\text{s}$$

For the circuit layouts shown in Figs 3.4(b) and (d), the component values can be derived from the relationships:

$$R_1/R_2 = 12.38 \qquad C_1.R_1 = 2937 \text{ } \mu\text{s} \qquad \text{and } C_2.R_2 = 81.1 \text{ } \mu\text{s}$$

The circuit arrangements shown in Figs 3.4(c) and (d), use 'shunt' type negative feedback (i.e., that type in which the negative feedback signal is applied to the amplifier in parallel with the input signal) connected around an internal gain block.

These layouts do not suffer from the same limitations in respect of source or load as the simple passive equalisation systems of (a) and (b). However, they do have the practical snag that the value of R_{in} will be determined by the required p.u. input load resistor (usually 47k. for a typical moving magnet or variable reluctance type of PU cartridge), and this sets an input 'resistor noise' threshold which is higher than desirable, as well as requiring inconveniently high values for R_1 and R_2.

For these reasons, the circuit arrangements shown in Figs 3.4(e) and (f), are much more commonly found in commercial audio circuitry. In these layouts, the frequency response shaping components are contained within a 'series' type feedback network (i.e., one in which the negative feedback signal is connected to the amplifier in series with the input signal), which means that the input circuit impedance seen by the amplifier is essentially that of the pick-up coil alone, and allows a lower mid-range 'thermal noise' background level.

The snag, in this case, is that at very high frequencies, where the impedance of the frequency-shaping feedback network is small in relation to R_{FB}, the circuit gain approaches unity, whereas both the RIAA specification and the accurate reproduction of transient waveforms require that the gain should asymptote to zero at higher audio frequencies.

This error in the shape of the upper half of the response curve can be remedied by the addition of a further *CR* network, C_3/R_3, on the output of the equalisation circuit, as shown in Figs 3.4(e) and (f). This amendment is sometimes found in the circuit designs used by the more perfectionist of the audio amplifier manufacturers.

Other approaches to the problem of combining low input noise levels with accurate replay equalisation are to divide the equalisation circuit into two parts, in which the first part, which can be based on a low noise series feedback layout, is only required to shape the 20 Hz−1 kHz section of the response curve. This can then be followed by either a simple passive *RC* roll-off network, as shown in Fig. 3.4(g), or by some other circuit arrangement having a similar effect − such as that based on the use of shunt feedback connected around an inverting amplifier stage, as shown in Fig. 3.4(h) – to generate that part of the response curve lying between 1 kHz and 20 kHz.

A further arrangement, which has attracted the interest of some Japanese circuit designers – as used, for example, in the Rotel RC-870BX preamp., of which the RIAA equalising circuit is shown in a simplified form in Fig. 3.4(j) – simply employs one of the recently developed very low noise IC op. amps as a flat frequency response input buffer stage. This is used to amplify the input signal to a level at which circuit noise introduced by succeeding stages will only be a minor problem, and also to convert the PU input impedance level to a value at which a straightforward shunt feedback equalising circuit can be used, with resistor values chosen to minimise any thermal noise background, rather than dictated by the PU load requirements.

The use of 'application specific' audio ICs, to reduce the cost and component count of RIAA stages and other circuit functions, has become much less popular among the designers of higher quality audio equipment because of the tendency of the semiconductor manufacturers to discontinue the supply of such specialised ICs when the economic basis of their sales becomes unsatisfactory, or to replace these devices by other, notionally equivalent, ICs which are not necessarily either pin or circuit function compatible.

There is now, however, a degree of unanimity among the suppliers of ICs as to the pin layout and operating conditions of the single and dual op. amp. designs, commonly packaged in 8-pin dual-in-line forms. These are typified by the Texas Instruments TL071 and TL072 ICs, or their more recent equivalents, such as the TL051 and TL052 devices – so there is a growing tendency for circuit designers to base their circuits on the use of ICs of this type, and it is assumed that devices of this kind would be used in the circuits shown in Fig. 3.4.

An incidental advantage of the choice of this style of IC is that commercial rivalry between semiconductor manufacturers leads to continuous improvements in the specification of these devices. Since these nearly always offer plug-in physical and electrical interchangeability, the performance of existing equipment can easily be up-graded, either on the production line or by the service department, by the replacement of existing op. amp. ICs with those of a more recent vintage, which is an advantage to both manufacturer and user.

MOVING COIL PU HEAD AMPLIFIER DESIGN

The design of pre-amplifier input circuitry which will accept the very low signal levels associated with moving coil pick-ups presents special problems in attaining an adequately high signal-to-noise ratio, in respect of the microvolt level input signals, and in minimising the intrusion of mains hum or unwanted RF signals.

The problem of circuit noise is lessened somewhat in respect of such RIAA equalised amplifier stages in that, because of the shape of the frequency response curve, the effective bandwidth of the amplifier is only about 800 Hz. The thermal noise due to the amplifier input impedance, which is defined by the equation below, is proportional to the squared measurement bandwidth, other things being equal, so the noise due to such a stage is less than would have been the case for a flat frequency response system, nevertheless, the attainment of an adequate S/N ratio, which should be at least 60 dB, demands that the input circuit impedance should not exceed some 50 ohms.

$$\bar{V} = \sqrt{4KT\,\delta FR}$$

where δF is the bandwidth, T is the absolute temperature, (room temperature being approx. 300° K), R is resistance in ohms and K is Boltzmann's constant (1.38×10^{-23}).

The moving coil pick-up cartridges themselves will normally have winding resistances which are only of the order of 5−25 ohms, except in the case of the high output units where the problem is less acute anyway, so the problem relates almost exclusively to the circuit impedance of the MC input circuitry and the semiconductor devices used in it.

CIRCUIT ARRANGEMENTS

Five different approaches are in common use for moving coil PU input amplification.

Step-up transformer

This was the earliest method to be explored, and was advocated by Ortofon, which was one of the pioneering companies in the manufacture of MC PU designs. The advantage of this system is that it is substantially noiseless, in the sense that the only source of wide-band noise will be the circuit impedance of the transformer windings, and that the output voltage can be high enough to minimise the thermal noise contribution from succeeding stages.

The principal disadvantages with transformer step-up systems, when these are operated at very low signal levels, are their proneness to mains 'hum' pick up, even when well shrouded, and their somewhat less good handling of 'transients', because of the effects of stray capacitances and leakage inductance. Care in their design is also needed to overcome the magnetic non-linearities associated with the core, which will be particularly significant at low signal levels.

Systems using paralleled input transistors

The need for a very low input circuit impedance to minimise thermal noise effects has been met in a number of commercial designs by simply connecting a number of small signal transistors in parallel to reduce their effective base-emitter circuit resistance. Designs of this type came from Ortofon, Linn/Naim, and Braithwaite, and are shown in Figs 3.5–3.7.

If such small signal transistors are used without selection and matching − a time-consuming and expensive process for any commercial manufac- urer − some means must be adopted to minimise the effects of the variation in base-emitter turn-on voltage which will exist between nomi- nally identical devices, due to variations in doping level in the silicon crystal slice, or to other differences in manufacture.

In the Ortofon circuit this is achieved by individual collector-base bias current networks, for which the penalty is the loss of some usable signal in the collector circuit. In the Linn/Naim and Braithwaite designs, this evening out of transistor characteristics in circuits having common base

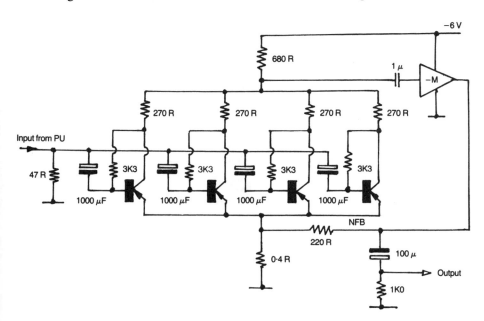

Fig. 3.5 *Ortofon MCA-76 head amplifier.*

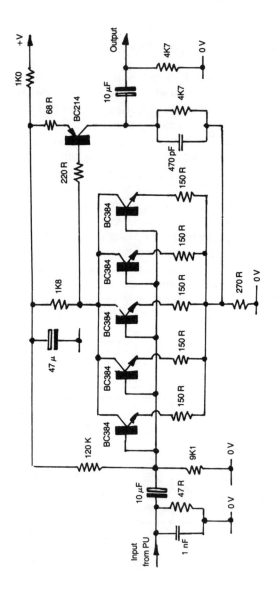

Fig. 3.6 *The Naim NAC 20 moving coil head amplifier.*

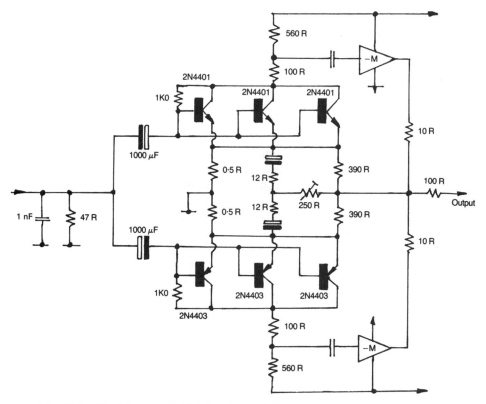

Fig. 3.7 *Braithwaite RA14 head amplifier. (Output stage shown in a simplified form.)*

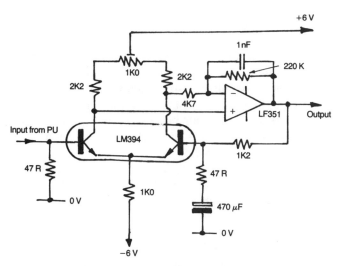

Fig. 3.8 *Head amplifier using LM394 multiple transistor array.*

connections is achieved by the use of individual emitter resistors to swamp such differences in device characteristics. In this case, the penalty is that such resistors add to the base-emitter circuit impedance, when the task of the design is to reduce this.

Monolithic super-matched input devices

An alternative method of reducing the input circuit impedance, without the need for separate bias systems or emitter circuit swamping resistors, is to employ a monolithic (integrated circuit type) device in which a multiplicity of transistors have been simultaneously formed on the same silicon chip. Since these can be assumed to have virtually identical characteristics they can be paralleled, at the time of manufacture, to give a very low impedance, low noise, matched pair.

An example of this approach is the National Semiconductors LM194/394 super-match pair, for which a suitable circuit is shown in Fig. 3.8. This input device probably offers the best input noise performance currently available, but is relatively expensive.

Small power transistors as input devices

The base-emitter impedance of a transistor depends largely on the size of the junction area on the silicon chip. This will be larger in power transistors than in small signal transistors, which mainly employ relatively small chip sizes. Unfortunately, the current gain of power transistors tends to decrease at low collector current levels, and this would make them unsuitable for this application.

However, the use of the plastic encapsulated medium power (3–4A Ic max.) styles, in T0126, T0127 and T0220 packages, at collector currents in the range 1–3 mA, achieves a satisfactory compromise between input circuit impedance and transistor performance, and allows the design of very linear low-noise circuitry. Two examples of MC head amplifier designs of this type, by the author, are shown in Figs 3.9 and 3.10.

The penalty in this case is that, because such transistor types are not specified for low noise operation, some preliminary selection of the devices is desirable, although, in the writer's experience, the bulk of the devices of the types shown will be found to be satisfactory in this respect.

In the circuit shown in Fig. 3.9, the input device is used in the common base (cascode) configuration, so that the input current generated by the pick-up cartridge is transferred directly to the higher impedance point at the collector of this transistor, so that the stage gain, prior to the application of negative feedback to the input transistor base, is simply the impedance transformation due to the input device.

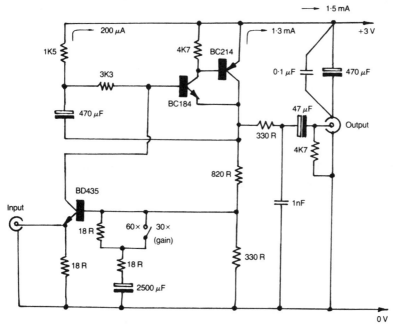

Fig. 3.9   *Cascode input moving coil head amplifier.*

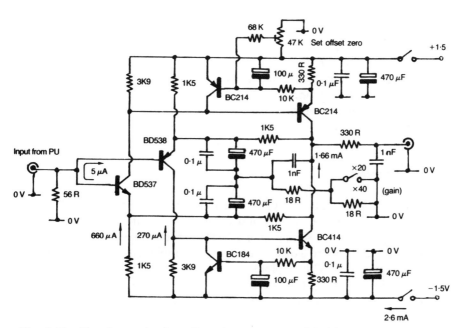

Fig. 3.10   *Very low-noise, low-distortion, symmetrical MC head amplifier.*

In the circuit of Fig. 3.10, the input transistors are used in a more conventional common-emitter mode, but the two input devices, though in a push-pull configuration, are effectively connected in parallel so far as the input impedance and noise figure are concerned. The very high degree of symmetry of this circuit assists in minimising both harmonic and transient distortions.

Both of these circuits are designed to operate from 3 V DC 'pen cell' battery supplies to avoid the introduction of mains hum due to the power supply circuitry or to earth loop effects. In mains-powered head amps. great care is always necessary to avoid supply line signal or noise intrusions, in view of the very low signal levels at both the inputs and the outputs of the amplifier stage.

It is also particularly advisable to design such amplifiers with single point '0 V' line and supply line connections, and these should be coupled by a suitable combination of good quality decoupling capacitors.

Very low noise IC op. amps

The development, some years ago, of very low noise IC operational amplifiers, such as the Precision Monolithics OP-27 and OP-37 devices, has led to the proliferation of very high quality, low-noise, low-distortion ICs aimed specifically at the audio market, such as the Signetics NE-5532/5534, the NS LM833, the PMI SSM2134/2139, and the TI TL051/052 devices.

With ICs of this type, it is a simple matter to design a conventional RIAA input stage in which the provision of a high sensitivity, low noise, moving coil PU input is accomplished by simply reducing the value of the input load resistor and increasing the gain of the RIAA stage in comparison with that needed for higher output PU types. An example of a typical Japanese design of this type is shown in Fig. 3.11.

Other approaches

A very ingenious, fully symmetrical circuit arrangement which allows the use of normal circuit layouts and components in ultra-low noise (e.g., moving coil p.u. and similar signal level) inputs, has been introduced by 'Quad' (Quad Electroacoustics Ltd) and is employed in all their current series of preamps. This exploits the fact that, at low input signal levels, bipolar junction transistors will operate quite satisfactorily with their base and collector junctions at the same DC potential, and permits the type of input circuit shown in Fig. 3.12.

In the particular circuit shown, that used in the 'Quad 44' disc input, a

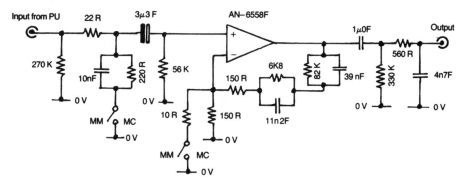

Fig. 3.11 *Moving coil/moving magnet RIAA input stage in Technics SU-V10 amplifier.*

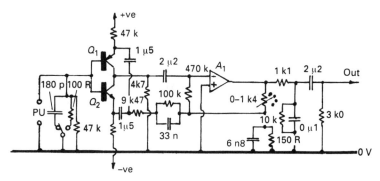

Fig. 3.12 *The 'Quad' ultra-low noise input circuit layout.*

two-stage equalisation layout is employed, using the type of structure illustrated in Fig. 3.4(g), with the gain of the second stage amplifier (a TL071 IC op. amp.) switchable to suit the type of input signal level available.

INPUT CONNECTIONS

For all low-level input signals care must be taken to ensure that the connections are of low contact resistance. This is obviously an important matter in the case of low-impedance circuits such as those associated with MC pick-up inputs, but is also important in higher impedance circuitry since the resistance characteristics of poor contacts are likely to be non-linear, and to introduce both noise and distortion.

In the better class modern units the input connectors will invariably be of the 'phono' type, and both the plugs and the connecting sockets will be gold plated to reduce the problem of poor connections due to contamination or tarnishing of the metallic contacts.

The use of separate connectors for L and R channels also lessens the problem of inter-channel breakthrough, due to capacitative coupling or leakage across the socket surface, a problem which can arise in the five- and seven-pin DIN connectors if they are carelessly fitted, and particularly when both inputs and outputs are taken to that same DIN connector.

INPUT SWITCHING

The comments made about input connections are equally true for the necessary switching of the input signal sources. Separate, but mechanically interlinked, switches of the push-on, push-off, type are to be preferred to the ubiquitous rotary wafer switch, in that it is much easier, with separate switching elements, to obtain the required degree of isolation between inputs and channels than would be the case when the wiring is crowded around the switch wafer.

However, even with separate push switches, the problem remains that the input connections will invariably be made to the rear of the amplifier/ preamplifier unit, whereas the switching function will be operated from the front panel, so that the internal connecting leads must traverse the whole width of the unit.

Other switching systems, based on relays, or bipolar or field effect transistors, have been introduced to lessen the unwanted signal intrusions which may arise on a lengthy connecting lead. The operation of a relay, which will behave simply as a remote switch when its coil is energised by a suitable DC supply, is straightforward, though for optimum performance it should either be hermetically sealed or have noble metal contacts to resist corrosion.

Transistor switching

Typical bipolar and FET input switching arrangements are shown in Figs 3.13 and 3.14. In the case of the bipolar transistor switch circuit of Fig. 3.13, the non-linearity of the junction device when conducting precludes its use in the signal line, the circuit is therefore arranged so that the transistor is non-conducting when the signal is passing through the con- trolled signal channel, but acts as a short-circuit to shunt the signal path to the O V line when it is caused to conduct.

In the case of the FET switch, if R_1 and R_2 are high enough, the non-

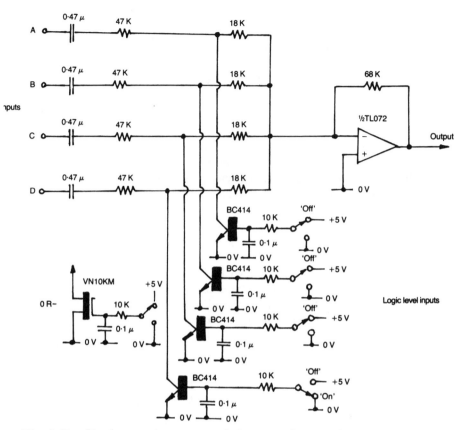

Fig. 3.13 *Bipolar transistor operated shunt switching. (Also suitable for small-power MOSFET devices.)*

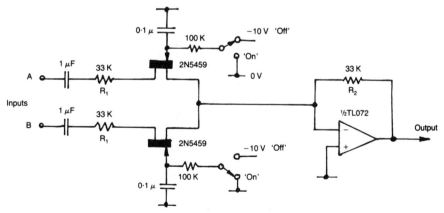

Fig. 3.14 *Junction FET input switching circuit.*

linearity of the conducting resistance of the FET channel will be swamped, and the harmonic and other distortions introduced by this device will be negligible. (Typically less than 0.02% at 1 V RMS and 1 kHz.)

The CMOS bilateral switches of the CD4066 type are somewhat non-linear, and have a relatively high level of breakthrough. For these reasons they are generally thought to be unsuitable for high quality audio equipment, where such remote switching is employed to minimise cross-talk and hum pick-up.

However, such switching devices could well offer advantages in lower quality equipment where the cost savings is being able to locate the switching element on the printed circuit board, at the point where it was required, might offset the device cost.

Diode switching

Diode switching of the form shown in Fig. 3.15, while very commonly employed in RF circuitry, is unsuitable for audio use because of the large shifts in DC level between the 'on' and 'off' conditions, and this would produce intolerable 'bangs' on operation.

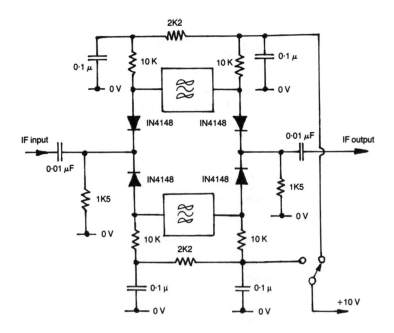

Fig. 3.15 *Typical diode switching circuit, as used in RF applications.*

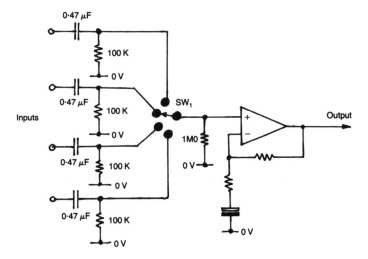

Fig. 3.16 *Use of DC blocking capacitors to minimise input switching noises.*

For all switching, quietness of operation is an essential requirement, and this demands that care shall be taken to ensure that all of the switched inputs are at the same DC potential, preferably that of the 0 V line. For this reason, it is customary to introduce DC blocking capacitors on all input lines, as shown in Fig. 3.16, and the time constants of the input RC networks should be chosen so that there is no unwanted loss of low frequency signals due to this cause.

CHAPTER 4

Voltage amplifiers and controls

PREAMPLIFIER STAGES

The popular concept of hi-fi attributes the major role in final sound quality to the audio power amplifier and the output devices or output configuration which it uses. Yet in reality the pre-amplifier system, used with the power amplifier, has at least as large an influence on the final sound quality as the power amplifier, and the design of the voltage gain stages within the pre- and power amplifiers is just as important as that of the power output stages. Moreover, it is developments in the design of such voltage amplifier stages which have allowed the continuing improvement in amplifier performance.

The developments in solid-state linear circuit technology which have occurred over the past thirty years seem to have been inspired in about equal measure by the needs of linear integrated circuits, and by the demands of high-quality audio systems, and engineers working in both of these fields have watched each other's progress and borrowed from each other's designs.

In general, the requirements for voltage gain stages in both audio amplifiers and integrated-circuit operational amplifiers are very similar. These are that they should be linear, which implies that they are free from waveform distortion over the required output signal range, have as high a stage gain as is practicable, have a wide AC bandwidth and a low noise level, and are capable of an adequate output voltage swing.

The performance improvements which have been made over this period have been due in part to the availability of new or improved types of semiconductor device, and in part to a growing understanding of the techniques for the circuit optimisation of device performance. It is the interrelation of these aspects of circuit design which is considered below.

LINEARITY

Bipolar transistors

In the case of a normal bipolar (NPN or PNP) silicon junction transistor, for which the chip cross-section and circuit symbol is shown in Fig. 4.1,

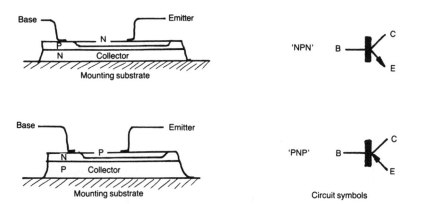

Fig. 4.1 *Typical chip cross-section of NPN and PNP silicon planar epitaxial transistors.*

the major problem in obtaining good linearity lies in the nature of the base voltage/collector current transfer characteristic, shown in the case of a typical 'NPN' device (a 'PNP' device would have a very similar characteristic, but with negative voltages and currents) in Fig. 4.2.

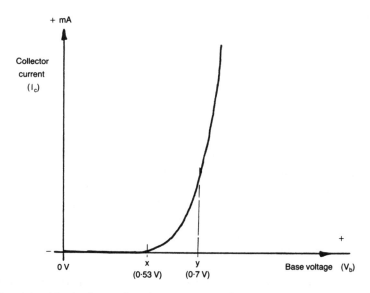

Fig. 4.2 *Typical transfer characteristic of silicon transistor.*

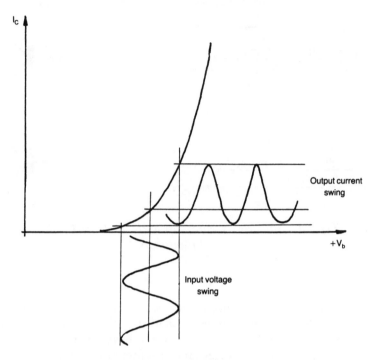

Fig. 4.3 *Transistor amplifier waveform distortion due to transfer characteristics.*

In this, it can be seen that the input/output transfer characteristic is strongly curved in the region 'X − Y' and an input signal applied to the base of such a device, which is forward biased to operate within this region, would suffer from the very prominent (second harmonic) waveform distortion shown in Fig. 4.3.

The way this type of non-linearity is influenced by the signal output level is shown in Fig. 4.4. It is normally found that the distortion increases as the output signal increases, and conversely.

There are two major improvements in the performance of such a bipolar amplifier stage which can be envisaged from these characteristics. Firstly, since the non-linearity is due to the curvature of the input characteristics of the device − the output characteristics, shown in Fig. 4.5, are linear − the smaller the input signal which is applied to such a stage, the lower the non-linearity, so that a higher stage gain will lead to reduced signal distortion at the same output level. Secondly, the distortion due to such a stage is very largely second harmonic in nature.

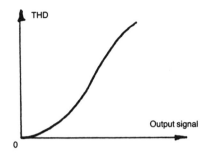

Fig. 4.4 *Relationship between signal distortion and output signal voltage in bipolar transistor amplifier.*

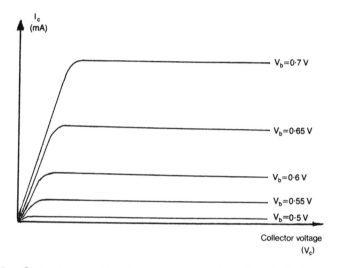

Fig. 4.5 *Output current/voltage characteristics of typical silicon bipolar transistor.*

This implies that a 'push-pull' arrangement, such as the so-called 'long-tailed pair' circuit shown in Fig. 4.6, which tends to cancel second harmonic distortion components, will greatly improve the distortion characteristics of such a stage.

Also, since the output voltage swing for a given input signal (the stage gain) will increase as the collector load (R_2 in Fig. 4.6) increases, the higher the effective impedance of this, the lower the distortion which will be introduced by the stage, for any given output voltage signal.

If a high value resistor is used as the collector load for Q_1 in Fig. 4.6, either a very high supply line voltage must be applied, which may exceed the voltage ratings of the devices or the collector current will be very

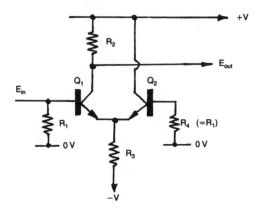

Fig. 4.6   *Transistor voltage amplifier using long-tailed pair circuit layout.*

small, which will reduce the gain of the device, and therefore tend to diminish the benefit arising from the use of a higher value load resistor.

Various circuit techniques have been evolved to circumvent this problem, by producing high dynamic impedance loads, which nevertheless permit the amplifying device to operate at an optimum value of collector current. These techniques will be discussed below.

An unavoidable problem associated with the use of high values of collector load impedance as a means of attaining high stage gains in such amplifier stages is that the effect of the 'stray' capacitances, shown as C_s in Fig. 4.7, is to cause the stage gain to decrease at high frequencies as the impedance of the stray capacitance decreases and progressively begins to shunt the load. This effect is shown in Fig. 4.8, in which the 'transition' frequency, f_o, (the -3 dB gain point) is that frequency at which the shunt impedance of the stray capacitance is equal to that of the load resistor, or its effective equivalent, if the circuit design is such that an 'active load' is used in its place.

Field effect devices

Other devices which may be used as amplifying components are field effect transistors and MOS devices. Both of these components are very much more linear in their transfer characteristics but have a very much lower mutual conductance (G_m).

This is a measure of the rate of change of output current as a function of an applied change in input voltage. For all bipolar devices, this is strongly dependent on collector current, and is, for a small signal silicon transistor, typically of the order of 45 mA/V, per mA collector current.

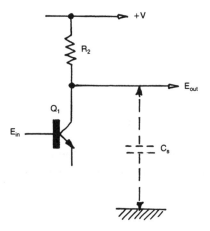

Fig. 4.7 *Circuit effect of stray capacitance.*

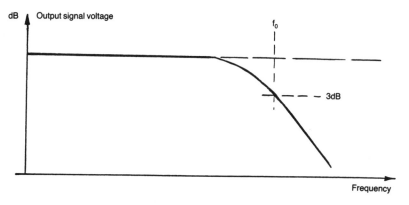

Fig. 4.8 *Influence of circuit stray capacitances on stage gain.*

Power transistors, operating at relatively high collector currents, for which a similar relationship applies, may therefore offer mutual conductances in the range of amperes/volt.

Since the output impedance of an emitter follower is approximately $1/G_m$, power output transistors used in this configuration can offer very low values of output impedance, even without externally applied negative feedback.

All field effect devices have very much lower values for G_m, which will lie, for small-signal components, in the range $2-10$ mA/V, not significantly affected by drain currents. This means that amplifier stages employing field effect transistors, though much more linear, offer much lower stage gains, other things being equal.

The transfer characteristics of junction (bipolar) FETs, and enhancement and depletion mode MOSFETS are shown in Figs 4.9(a), (b) and (c).

MOSFETs

MOSFETs, in which the gate electrode is isolated from the source/drain channel, have very similar transfer characteristics to that of junction FETs. They have an advantage that, since the gate is isolated from the drain/source channel by a layer of insulation, usually silicon oxide or nitride, there is no maximum forward gate voltage which can be applied — within the voltage breakdown limits of the insulating layer. In a junction FET the gate, which is simply a reverse biassed PN diode junction, will conduct if a forward voltage somewhat in excess of 0.6 V is applied.

The chip constructions and circuit symbols employed for small signal lateral MOSFETs and junction FETs (known simply as FETs) are shown in Figs 4.10 and 4.11.

It is often found that the chip construction employed for junction FETs is symmetrical, so that the source and drain are interchangeable in use.

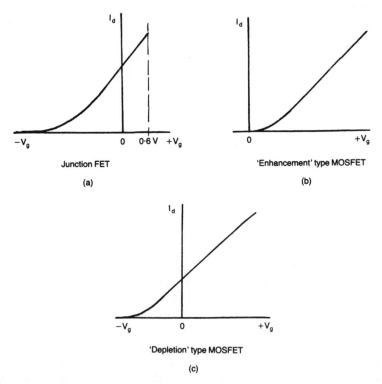

Fig. 4.9 *Gate voltage versus drain current characteristics of field effect devices.*

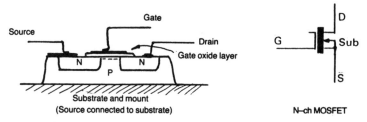

Fig. 4.10 *Chip cross-section and circuit symbol for lateral MOSFET (small signal type)*

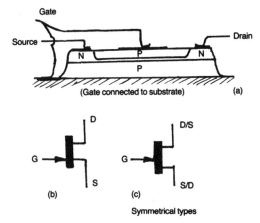

Fig. 4.11 *Chip cross-section and circuit symbols for (bipolar) junction FET.*

For such devices the circuit symbol shown in Fig. 4.11(c) should properly be used.

A practical problem with lateral devices, in which the current flow through the device is parallel to the surface of the chip, is that the path length from source to drain, and hence the device impedance and current carrying capacity, is limited by the practical problems of defining and etching separate regions which are in close proximity, during the manufacture of the device.

V-MOS AND T-MOS
This problem is not of very great importance for small signal devices, but it is a major concern in high current ones such as those employed in power output stages. It has led to the development of MOSFETs in which the current flow is substantially in a direction which is vertical to the surface, and in which the separation between layers is determined by diffusion processes rather than by photo-lithographic means.

Devices of this kind, known as V-MOS and T-MOS constructions, are shown in Figs 4.12(a) and (b).

Although these were originally introduced for power output stages, the electrical characteristics of such components are so good that these have been introduced, in smaller power versions, specifically for use in small signal linear amplifier stages. Their major advantages over bipolar devices, having equivalent chip sizes and dissipation ratings, are their high input impedance, their greater linearity, and their freedom from 'hole storage' effects if driven into saturation.

These qualities are increasingly attracting the attention of circuit designers working in the audio field, where there is a trend towards the design of amplifiers having a very high intrinsic linearity, rather than relying on the use of negative feedback to linearise an otherwise worse design.

BREAKDOWN

A specific problem which arises in small signal MOSFET devices is that, because the gate-source capacitance is very small, it is possible to induce breakdown of the insulating layer, which destroys the device, as a result of transferred static electrical charges arising from mishandling.

Though widely publicised and the source of much apprehension, this

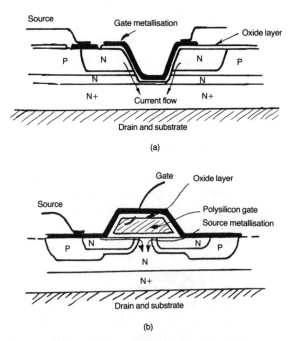

Fig. 4.12 *Power MOSFET constructions using (a) V and (b) T configurations. (Practical devices will employ many such cells in parallel.)*

problem is actually very rarely encountered in use, since small signal MOSFETs usually incorporate protective zener diodes to prevent this eventuality, and power MOSFETs, where such diodes may not be used because they may lead to inadvertent 'thyristor' action, have such a high gate-source capacitance that this problem does not normally arise.

In fact, when such power MOSFETs do fail, it is usually found to be due to circuit design defects, which have either allowed excessive operating potentials to be applied to the device, or have permitted inadvertent VHF oscillation, which has led to thermal failure.

NOISE LEVELS

Improved manufacturing techniques have lessened the differences between the various types of semiconductor device, in respect of intrinsic noise level. For most practical purposes it can now be assumed that the characteristics of the device will be defined by the thermal noise figure of the circuit impedances. This relationship is shown in the graph of Fig. 4.13.

For very low noise systems, operating at circuit impedance levels which have been deliberately chosen to be as low as practicable – such as in moving coil pick-up head amplifiers – bipolar junction transistors are still the preferred device. These will either be chosen to have a large base junction area, or will be employed as a parallel-connected array; as, for example, in the LM194/394 'super-match pair' ICs, where a multiplicity of parallel connected transistors are fabricated on a single chip, giving an effective input (noise) impedance as low as 40 ohms.

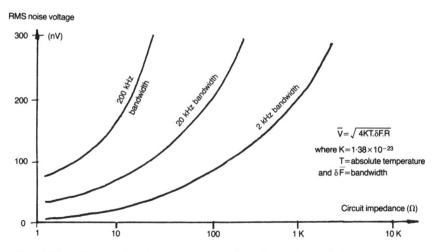

Fig. 4.13 *Thermal noise output as a function of circuit impedance.*

However, recent designs of monolithic dual J-FETs, using a similar type of multiple parallel-connection system, such as the Hitachi 2SK389, can offer equivalent thermal noise resistance values as low as 33 ohms, and a superior overall noise figure at input resistance values in excess of 100 ohms.

At impedance levels beyond about 1 kilohm there is little practical difference between any devices of recent design. Earlier MOSFET types were not so satisfactory, due to excess noise effects arising from carrier trapping mechanisms in impurities at the channel/gate interface.

OUTPUT VOLTAGE CHARACTERISTICS

Since it is desirable that output overload and signal clipping do not occur in audio systems, particularly in stages preceding the gain controls, much emphasis has been placed on the so-called 'headroom' of signal handling stages, especially in hi-fi publications where the reviewers are able to distance themselves from the practical problems of circuit design.

While it is obviously desirable that inadvertent overload shall not occur in stages preceding signal level controls, high levels of feasible output voltage swing demand the use of high voltage supply rails, and this, in turn, demands the use of active components which can support such working voltage levels.

Not only are such devices more costly, but they will usually have poorer performance characteristics than similar devices of lower voltage ratings. Also, the requirement for the use of high voltage operation may preclude the use of components having valuable characteristics, but which are restricted to lower voltage operation.

Practical audio circuit designs will therefore regard headroom simply as one of a group of desirable parameters in a working system, whose design will be based on careful consideration of the maximum input signal levels likely to be found in practice.

Nevertheless, improved transistor or IC types, and new developments in circuit architecture, are welcomed as they occur, and have eased the task of the audio design engineer, for whom the advent of new programme sources, in particular the compact disc, and now digital audio tape systems, has greatly extended the likely dynamic range of the output signal.

Signal characteristics

The practical implications of this can be seen from a consideration of the signal characteristics of existing programme sources. Of these, in the past, the standard vinyl ('black') disc has been the major determining factor. In this, practical considerations of groove tracking have limited the recorded needle tip velocity to about 40 cm/s, and typical high-quality pick-up cartridges capable of tracking this recorded velocity will have a voltage output of some 3 mV at a standard 5 cm/s recording level.

If the pre-amplifier specification calls for maximum output to be obtainable at a 5 cm/s input, then the design should be chosen so that there is a 'headroom factor' of at least 8×, in such stages preceding the gain controls.

In general, neither FM broadcasts, where the dynamic range of the transmitted signal is limited by the economics of transmitter power, nor cassette recorders, where the dynamic range is constrained by the limited tape overload characteristics, have offered such a high practicable dynamic range.

It is undeniable that the analogue tape recorder, when used at 15 in./s, twin-track, will exceed the LP record in dynamic range. After all, such recorders were originally used for mastering the discs. But such programme sources are rarely found except among 'live recording' enthusiasts. However, the compact disc, which is becoming increasingly common among purely domestic hi-fi systems, presents a new challenge, since the practicable dynamic range of this system exceeds 80 dB (10 000:1), and the likely range from mean (average listening level) to peak may well be as high as 35 dB (56:1) in comparison with the 18 dB (8:1) range likely with the vinyl disc.

Fortunately, since the output of the compact disc player is at a high level, typically 2 V RMS, and requires no signal or frequency response conditioning prior to use, the gain control can be sited directly at the input of the preamp. Nevertheless, this still leaves the possibility that signal peaks may occur during use which are some 56× greater than the mean programme level, with the consequence of the following amplifier stages being driven hard into overload.

This has refocused attention on the design of solid state voltage amplifier stages having a high possible output voltage swing, and upon power amplifiers which either have very high peak output power ratings, or more graceful overload responses.

VOLTAGE AMPLIFIER DESIGN

The sources of non-linearity in bipolar junction transistors have already been referred to, in respect of the influence of collector load impedance, and push–pull symmetry in reducing harmonic distortion. An additional factor with bipolar junction devices is the external impedance in the base circuit, since the principal non-linearity in a bipolar device is that due to its input voltage/output current characteristics. If the device is driven from a high impedance source, its linearity will be substantially greater, since it is operating under conditions of current drive.

This leads to the good relative performance of the simple, two-stage, bipolar transistor circuit of Fig. 4.14, in that the input transistor, Q_1, is only required to deliver a very small voltage drive signal to the base of Q_2, so the signal distortion due to Q_1 will be low. Q_2, however, which is required to develop a much larger output voltage swing, with a much greater potential signal non-linearity, is driven from a relatively high source impedance, composed of the output impedance of Q_1, which is very high indeed, in parallel with the base-emitter resistor, R_4. R_1, R_2, and R_3/C_2 are employed to stabilise the DC working conditions of the circuit.

Normally, this circuit is elaborated somewhat to include both DC and AC negative feedback from the collector of Q_2 to the emitter of Q_1, as shown in the practical amplifier circuit of Fig. 4.15.

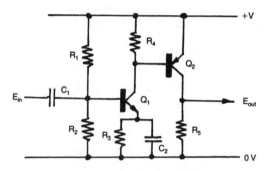

Fig. 4.14 *Two-stage transistor voltage amplifier.*

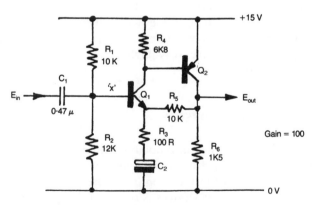

Fig. 4.15 *Practical two-transistor feedback amplifier.*

This is capable of delivering a 14 V p-p output swing, at a gain of 100, and a bandwidth of 15 Hz to 250 kHz, at −3 dB points; largely determined by the value of C_2 and the output capacitances, with a THD figure of better that 0.01% at 1 kHz.

The practical drawbacks of this circuit relate to the relatively low value necessary for R_3 − with the consequent large value necessary for C_2 if a good LF response is desired, and the DC offset between point 'X' and the output, due to the base-emitter junction potential of Q_1, and the DC voltage drop along R_5, which makes this circuit relatively unsuitable in DC amplifier applications.

An improved version of this simple two-stage amplifier circuit is shown in Fig. 4.16, in which the single input transistor has been replaced by a 'long-tailed pair' configuration of the type shown in Fig. 4.16. In this, if the two-input transistors are reasonably well matched in current gain, and if the value of R_3 is chosen to give an equal collector current flow through both Q_1 and Q_2, the DC offset between input and output will be negligible, and this will allow the circuit to be operated between symmetrical (+ and −) supply rails, over a frequency range extending from DC to 250 kHz or more.

Because of the improved rejection of odd harmonic distortion inherent in the input 'push-pull' layout, the THD due to this circuit, particularly at less than maximum output voltage swing, can be extremely low, and this probably forms the basis of the bulk of linear amplifier designs. However, further technical improvements are possible, and these are discussed below.

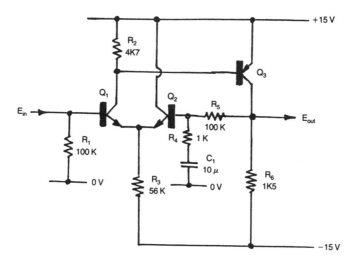

Fig. 4.16 *Improved two-stage feedback amplifier.*

CONSTANT-CURRENT SOURCES AND 'CURRENT MIRRORS'

As mentioned above, the use of high-value collector load resistors in the interests of high stage gain and low inherent distortion carries with it the penalty that the performance of the amplifying device may be impaired by the low collector current levels which result from this approach.

Advantage can, however, be taken of the very high output impedance of a junction transistor, which is inherent in the type of collector current/supply voltage characteristics illustrated in Fig. 4.5, where even at currents in the $1-10$ mA region, dynamic impedances of the order of 100 kilohms may be expected.

A typical circuit layout which utilises this characteristic is shown in Fig. 4.17, in which R_1 and R_2 form a potential divider to define the base potential of Q_1, and R_3 defines the total emitter or collector currents for this effective base potential.

This configuration can be employed with transistors of either PNP or NPN types, which allows the circuit designer considerable freedom in their application.

An improved, two-transistor, constant current source is shown in Fig. 4.18. In this R_1 is used to bias Q_2 into conduction, and Q_1 is employed to sense the voltage developed across R_2, which is proportional to emitter current, and to withdraw the forward bias from Q_2 when that current level is reached at which the potential developed across R_2 is just sufficient to cause Q_1 to conduct.

The performance of this circuit is greatly superior to that of Fig. 4.17, in that the output impedance is about 10× greater, and the circuit is insensitive to the potential, $+V_{ref.}$, applied to R_1, so long as it is adequate to force both Q_2 and Q_1 into conduction.

An even simpler circuit configuration makes use of the inherent very high output impedance of a junction FET under constant gate bias conditions. This employs the circuit layout shown in Fig. 4.19, which allows a

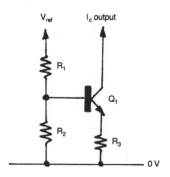

Fig. 4.17 *Transistor constant current source.*

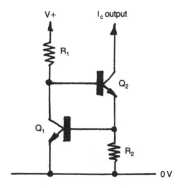

Fig. 4.18 *Two-transistor constant current source.*

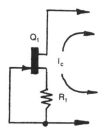

Fig. 4.19 *Two-terminal constant current source.*

true 'two-terminal' constant current source, independent of supply lines or reference potentials, and which can be used at either end of the load chain.

The current output from this type of circuit is controlled by the value chosen for R_1, and this type of constant current source may be constructed using almost any available junction FET, provided that the voltage drop across the FET drain-gate junction does not exceed the breakdown voltage of the device. This type of constant current source is also available as small, plastic-encapsulated, two-lead devices, at a relatively low cost, and with a range of specified output currents.

All of these constant current circuit layouts share the common small disadvantage that they will not perform very well at low voltages across the current source element. In the case of Figs 4.17 and 4.18, the lowest practicable operating potential will be about 1 V. The circuit of Fig. 4.19 may require, perhaps, 2–3 V, and this factor must be considered in circuit performance calculations.

The 'boot-strapped' load resistor arrangement shown in Fig. 4.20, and commonly used in earlier designs of audio amplifier to improve the

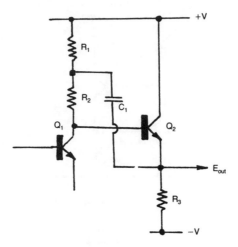

Fig. 4.20   *Load impedance increase by boot-strap circuit.*

linearity of the last class 'A' amplifier stage (Q_1), effectively multiplies the resistance value of R_2 by the gain which Q_2 would be deemed to have if operated as a common-emitter amplifier with a collector load of R_3 in parallel with R_1.

This arrangement is the best configuration practicable in terms of available RMS output voltage swing, as compared with conventional constant current sources, but has fallen into disuse because it leads to slightly less good THD figures than are possible with other circuit arrangements.

All these circuit arrangements suffer from a further disadvantage, from the point of view of the integrated circuit designer: they employ resistors as part of the circuit design, and resistors, though possible to fabricate in IC structures, occupy a disproportionately large area of the chip surface. Also, they are difficult to fabricate to precise resistance values without resorting to subsequent laser trimming, which is expensive and time-consuming.

Because of this, there is a marked preference on the part of IC design engineers for the use of circuit layouts known as 'current mirrors', of which a simple form is shown in Fig. 4.21.

IC solutions

These are not true constant current sources, in that they are only capable of generating an output current (I_{out}) which is a close equivalence of the input or drive current (I_{in}). However, the output impedance is very high, and if the drive current is held to a constant value, the output current will also remain constant.

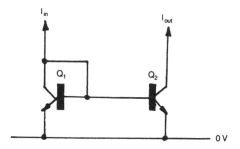

Fig. 4.21   *Simple form of current mirror.*

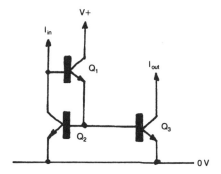

Fig. 4.22   *Improved form of current mirror.*

A frequently found elaboration of this circuit, which offers improvements in respect of output impedance and the closeness of equivalence of the drive and output currents, is shown in Fig. 4.22. Like the junction FET based constant current source, these current mirror devices are available as discrete, plastic-encapsulated, three-lead components, having various drive current/output current ratios, for incorporation into discrete component circuit designs.

The simple amplifier circuit of Fig. 4.16 can be elaborated, as shown in Fig. 4.23, to employ these additional circuit refinements, which would have the effect of increasing the open-loop gain, i.e. that before negative feedback is applied, by $10-100\times$ and improving the harmonic and other distortions, and the effective bandwidth by perhaps $3-10\times$. From the point of view of the IC designer, there is also the advantage of a potential reduction in the total resistor count.

These techniques for improving the performance of semiconductor amplifier stages find particular application in the case of circuit layouts employing junction FETs and MOSFETs, where the lower effective mutual conductance values for the devices would normally result in relatively poor stage gain figures.

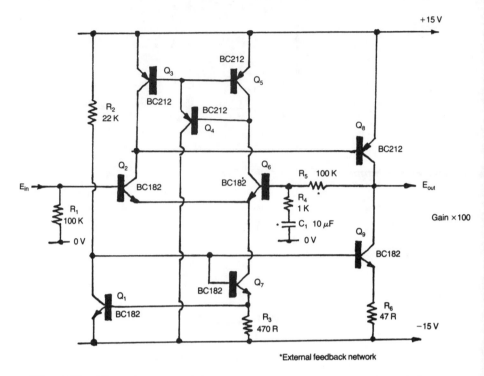

Fig. 4.23 *Use of circuit elaboration to improve two-stage amplifier of Fig. 4.16.*

This has allowed the design of IC operational amplifiers, such as the RCA CA3140 series, or the Texas Instruments TL071 series, which employ, respectively, MOSFET and junction FET input devices. The circuit layout employed in the TL071 is shown, by way of example, in Fig. 4.24.

Both of these op. amp. designs offer input impedances in the million megohm range − in comparison with the input impedance figures of 5−10 kilohm which were typical of early bipolar ICs − and the fact that the input impedance is so high allows the use of such ICs in circuit configurations for which earlier op. amp. ICs were entirely inappropriate.

Although the RCA design employs MOSFET input devices which offer, in principle, an input impedance which is perhaps 1000 times better than this figure, the presence of on-chip Zener diodes, to protect the device against damage through misuse or static electric charges, reduces the input impedance to roughly the same level as that of the junction FET device.

It is a matter for some regret that the design of the CA3140 series devices is now so elderly that the internal MOSFET devices do not offer

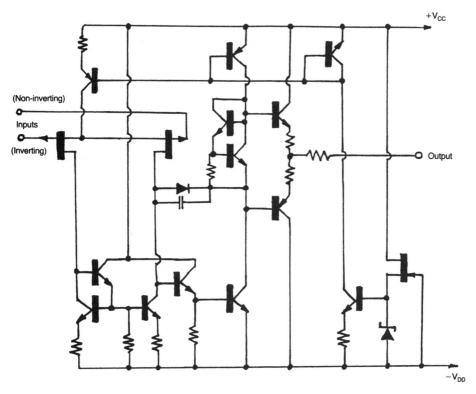

Fig. 4.24 *Circuit layout of Texas Instruments TL071 op. amp.*

the low level of internal noise of which more modern MOSFET types are capable. This tends rather to rule out the use of this MOSFET op. amp. for high quality audio use, though the TL071 and its equivalents such as the LF351 have demonstrated impeccable audio behaviour.

PERFORMANCE STANDARDS

It has always been accepted in the past, and is still held as axiomatic among a very large section of the engineering community, that performance characteristics can be measured, and that improved levels of measured performance will correlate precisely, within the ability of the ear to detect such small differences, with improvements which the listener will hear in reproduced sound quality.

Within a strictly engineering context, it is difficult to do anything other than accept the concept that measured improvements in performance are the only things which should concern the designer.

However, the frequently repeated claim by journalists and reviewers working for periodicals in the hi-fi field — who, admittedly, are unlikely to be unbiased witnesses — that measured improvements in performance do not always go hand-in-hand with the impressions that the listener may form, tends to undermine the confidence of the circuit designer that the instrumentally determined performance parameters are all that matter.

It is clear that it is essential for engineering progress that circuit design improvements must be sought which lead to measurable performance improvements. Yet there is now also the more difficult criterion that those things which appear to be better, in respect of measured parameters, must also be seen, or heard, to be better.

Use of ICs

This point is particularly relevant to the question of whether, in very high quality audio equipment, it is acceptable to use IC operational amplifiers, such as the TL071, or some of the even more exotic later developments such as the NE5534 or the OP27, as the basic gain blocks, around which the passive circuitry can be arranged, or whether, as some designers believe, it is preferable to construct such gain blocks entirely from discrete components.

Some years ago, there was a valid technical justification for this reluctance to use op. amp. ICs in high quality audio circuitry, since the method of construction of such ICs was as shown, schematically, in Fig. 4.25, in which all the structural components were formed on the surface of a heavily 'P' doped silicon substrate, and relied for their isolation, from one another or from the common substrate, on the reverse biased diodes formed between these elements.

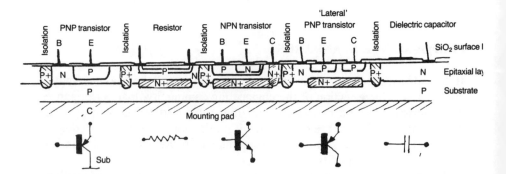

Fig. 4.25 *Method of fabrication of components in silicon integrated circuit.*

This led to a relatively high residual background noise level, in comparison with discrete component circuitry, due to the effects of the multiplicity of reverse diode leakage currents associated with every component on the chip. Additionally, there were quality constraints in respect of the components formed on the chip surface − more severe for some component types than for others − which also impaired the circuit performance.

A particular instance of this problem arose in the case of PNP transistors used in normal ICs, where the circuit layout did not allow these to be formed with the substrate acting as the collector junction. In this case, it was necessary to employ the type of construction known as a 'lateral PNP', in which all the junctions are diffused in, from the exposed chip surface, side by side.

In this type of device the width of the 'N' type base region, which must be very small for optimum results, depends mainly on the precision with which the various diffusion masking layers can be applied. The results are seldom very satisfactory. Such a lateral PNP device has a very poor current gain and HF performance.

In recent IC designs, considerable ingenuity has been shown in the choice of circuit layout to avoid the need to employ such unsatisfactory components in areas where their shortcomings would affect the end result. Substantial improvements, both in the purity of the base materials and in diffusion technology, have allowed the inherent noise background to be reduced to a level where it is no longer of practical concern.

Modern standards

The standard of performance which is now obtainable in audio applications, from some of the recent IC op. amps − especially at relatively low closed-loop gain levels − is frequently of the same order as that of the best discrete component designs, but with considerable advantages in other respects, such as cost, reliability and small size.

This has led to their increasing acceptance as practical gain blocks, even in very high quality audio equipment.

When blanket criticism is made of the use of ICs in audio circuitry, it should be remembered that the 741 which was one of the earliest of these ICs to offer a satisfactory performance − though it is outclassed by more recent types − has been adopted with enthusiasm, as a universal gain block, for the signal handling chains in many recording and broadcasting studios.

This implies that the bulk of the programme signals employed by the critics to judge whether or not a discrete component circuit is better than that using an IC, will already have passed through a sizeable handful of

741-based circuit blocks, and if such ICs introduce audible defects, then their reference source is already suspect.

It is difficult to stipulate the level of performance which will be adequate in a high-quality audio installation. This arises partly because there is little agreement between engineers and circuit designers on the one hand, and the hi-fi fraternity on the other, about the characteristics which should be sought, and partly because of the wide differences which exist between listeners in their expectations for sound quality or their sensitivity to distortions. These differences combine to make it a difficult and specu-lative task to attempt either to quantify or to specify the technical com-ponents of audio quality, or to establish an acceptable minimum quality level.

Because of this uncertainty, the designer of equipment, in which price is not a major consideration, will normally seek to attain standards substantially in excess of those which he supposes to be necessary, simply in order not to fall short. This means that the reason for the small residual differences in the sound quality, as between high quality units, is the existence of malfunctions of types which are not currently known or measured.

AUDIBILITY OF DISTORTION

Harmonic and intermodulation distortions

Because of the small dissipations which are normally involved, almost all discrete component voltage amplifier circuitry will operate in class 'A' (that condition in which the bias applied to the amplifying device is such as to make it operate in the middle of the linear region of its input/output transfer characteristic), and the residual harmonic components are likely to be mainly either second or third order, which are audibly much more tolerable than higher order distortion components.

Experiments in the late 1940s suggested that the level of audibility for second and third harmonics was of the order of 0.6% and 0.25% respect-ively, and this led to the setting of a target value, within the audio spectrum, of 0.1% THD, as desirable for high quality audio equipment.

However, recent work aimed at discovering the ability of an average listener to detect the presence of low order (i.e. second or third) harmonic distortions has drawn the uncomfortable conclusion that listeners, taken from a cross section of the public, may rate a signal to which 0.5% second harmonic distortion has been added as 'more musical' than, and therefore preferable to, the original undistorted input. This discovery tends to cast doubt on the value of some subjective testing of equipment.

What is not in dispute is that the inter-modulation distortion, (IMD), which is associated with any non-linearity in the transfer characteristics, leads to a muddling of the sound picture, so that if the listener is asked, not which sound he prefers, but which sound seems to him to be the clearer, he will generally choose that with the lower harmonic content.

The way in which IMD arises is shown in Fig. 4.26, where a composite signal containing both high-frequency and low-frequency components, fed through a non-linear system, causes each signal to be modulated by the other. This is conspicuous in the drawing in respect of the HF component, but is also true for the LF one.

This can be shown mathematically to be due to the generation of sum and difference products, in addition to the original signal components, and provides a simple method, shown schematically in Fig. 4.27, for the

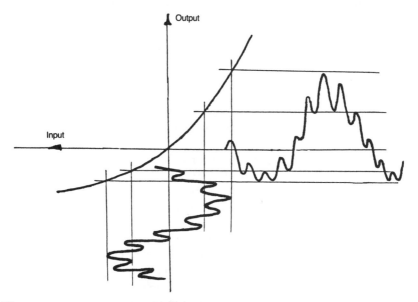

Fig. 4.26 *Inter-modulation distortions produced by the effect of a non-linear input/output transfer characteristic on a complex tone.*

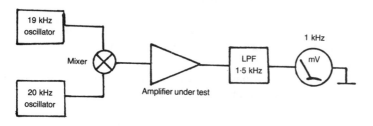

Fig. 4.27 *Simple HF two-tone inter-modulation distortion test.*

detection of this type of defect. A more formal IMD measurement system is shown in Fig. 4.28.

With present circuit technology and device types, it is customary to design for total harmonic and IM distortions to be below 0.01% over the range 30 Hz–20 kHz, and at all signal levels below the onset of clipping. Linear IC op. amps, such as the TL071 and the LF351, will also meet this spec. over the frequency range 30 Hz–10 kHz.

Transient defects

A more insidious group of signal distortions may occur when brief signals of a transient nature, or sudden step type changes in base level, are superimposed on the more continuous components of the programme signal. These defects can take the form of slew-rate distortions, usually associated with loss of signal during the period of the slew-rate saturation of the amplifier – often referred to as transient inter-modulation distortion or TID.

This defect is illustrated in Fig. 4.29, and arises particularly in amplifier systems employing substantial amounts of negative feedback, when there is some slew-rate limiting component within the amplifier, as shown in Fig. 4.30.

A further problem is that due to 'overshoot', or 'ringing', on a transient input, as illustrated in Fig. 4.31. This arises particularly in feedback amplifiers if there is an inadequate stability margin in the feedback loop, particularly under reactive load conditions, but will also occur in low-pass filter systems if too high an attenuation rate is employed.

The ear is very sensitive to slew-rate induced distortion, which is perceived as a 'tizziness' in the reproduced sound. Transient overshoot is normally noted as a somewhat over-bright quality. The avoidance of both these problems demands care in the circuit design, particularly when a constant current source is used, as shown in Fig. 4.32.

In this circuit, the constant current source, CC_1, will impose an absolute limit on the possible rate of change of potential across the capacitance C_1, (which could well be simply the circuit stray capacitance), when the output voltage is caused to move in a positive-going direction. This problem is compounded if an additional current limit mechanism, CC_2, is included in the circuitry to protect the amplifier transistor (Q_1) from output current overload.

Since output load and other inadvertent capacitances are unavoidable, it is essential to ensure that all such current limited stages operate at a current level which allows potential slewing to occur at rates which are at least 10× greater than the fastest signal components. Alternatively, means may be taken, by way of a simple input integrating circuit, (R_1C_1), as

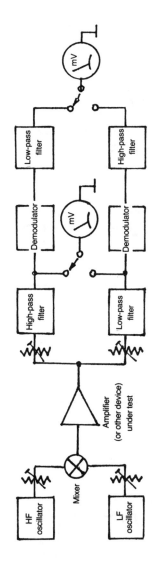

Fig. 4.28 *Two-tone inter-modulation distortion test rig.*

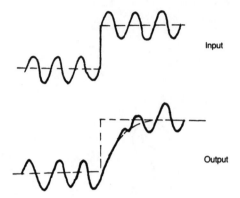

Fig. 4.29 *Effect of amplifier slew-rate saturation or transient inter-modulation distortion.*

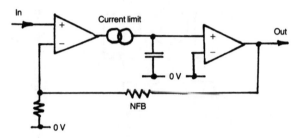

Fig. 4.30 *Typical amplifier layout causing slew-rate saturation.*

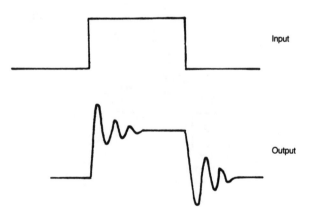

Fig. 4.31 *Transient 'ringing'.*

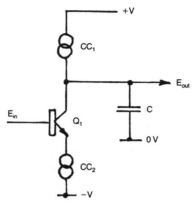

Fig. 4.32   *Circuit design aspects which may cause slew-rate limiting.*

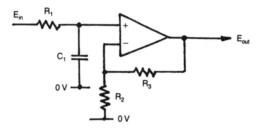

Fig. 4.33   *Input HF limiting circuit to lessen slew-rate limiting.*

shown in Fig. 4.33, to ensure that the maximum rate of change of the input signal voltage is within the ability of the amplifier to handle it.

Spurious signals

In addition to harmonic, IM, and transient defects in the signal channel, which will show up on normal instrumental testing, there is a whole range of spurious signals which may not arise in such tests. The most common of these is that of the intrusion of noise and alien signals, either from the supply line, or by direct radio pick-up.

This latter case is a random and capricious problem which can only be solved by steps appropriate to the circuit design in question. However, supply line intrusions, whether due to unwanted signals from the power supply, or from the other channel in a stereo system, may be greatly reduced by the use of circuit designs offering a high immunity to voltage fluctuations on the DC supply.

Other steps, such as the use of electronically stabilised DC supplies, or the use of separate power supplies in a stereo amplifier, are helpful, but the required high level of supply line signal rejection should be sought as a design feature before other palliatives are applied. Modern IC op. amps offer a typical supply voltage rejection ratio of 90 dB (30 000:1). Good discrete component designs should offer at least 80 dB (10 000:1).

This figure tends to degrade at higher frequencies, and this has led to the growing use of supply line bypass capacitors having a low effective series resistance (ESR). This feature is either a result of the capacitor design, or is achieved in the circuit by the designer's adoption of groups of parallel connected capacitors chosen so that the AC impedance remains low over a wide range of frequencies.

A particular problem in respect of spurious signals, which occurs in audio power amplifiers, is due to the loudspeaker acting as a voltage generator, when stimulated by pressure waves within the cabinet, and injecting unwanted audio components directly into the amplifier's negative feedback loop. This specific problem is unlikely to arise in small signal circuitry, but the designer must consider what effect output/line load characteristics may have – particularly in respect of reduced stability margin in a feedback amplifier.

In all amplifier systems there is a likelihood of microphonic effects due to the vibration of the components. This is likely to be of increasing importance at the input of 'low level', high sensitivity, pre-amplifier stages, and can lead to coloration of the signal when the equipment is in use, which is overlooked in the laboratory in a quiet environment.

Mains-borne interference

Mains-borne interference, as evidenced by noise pulses on switching electrical loads, is most commonly due to radio pick-up problems, and is soluble by the techniques (attention to signal and earth line paths, avoidance of excessive HF bandwidth at the input stages) which are applicable to these.

GENERAL DESIGN CONSIDERATIONS

During the past three decades, a range of circuit design techniques has been evolved to allow the construction of highly linear gain stages based on bipolar transistors whose input characteristics are, in themselves, very non-linear. These techniques have also allowed substantial improvements in possible stage gain, and have led to greatly improved performance from linear, but low gain, field-effect devices.

These techniques are used in both discrete component designs and in their monolithic integrated circuit equivalents, although, in general, the circuit designs employed in linear ICs are considerably more complex than those used in discrete component layouts.

This is partly dictated by economic considerations, partly by the requirements of reliability, and partly because of the nature of IC design.

The first two of these factors arise because both the manufacturing costs and the probability of failure in a discrete component design are directly proportional to the number of components used, so the fewer the better, whereas in an IC, both the reliability and the expense of manufacture are only minimally affected by the number of circuit elements employed.

In the manufacture of ICs, as has been indicated above, some of the components which must be employed are much worse than their discrete design equivalents. This has led the IC designer to employ fairly elaborate circuit structures, either to avoid the need to use a poor quality component in a critical position, or to compensate for its shortcomings.

Nevertheless, the ingenuity of the designers, and the competitive pressures of the market place, have resulted in systems having a very high performance, usually limited only by their inability to accept differential supply line potentials in excess of 36 V, unless non-standard diffusion processes are employed.

For circuitry requiring higher output or input voltage swings than allowed by small signal ICs, the discrete component circuit layout is, at the moment, unchallenged. However, as every designer knows, it is a difficult matter to translate a design which is satisfactory at a low working voltage design into an equally good higher voltage system.

This is because:

- increased applied potentials produce higher thermal dissipations in the components, for the same operating currents;
- device performance tends to deteriorate at higher inter-electrode potentials and higher output voltage excursions;
- available high/voltage transistors tend to be more restricted in variety and less good in performance than lower voltage types.

CONTROLS

These fall into a variety of categories:

- gain controls needed to adjust the signal level between source and power amplifier stages
- tone controls used to modify the tonal characteristics of the signal chain

- filters employed to remove unwanted parts of the incoming signal, and those adjustments used to alter the quality of the audio presentation, such as stereo channel balance or channel separation controls.

Gain controls

These are the simplest in basic form, and are often just a resistive potentiometer voltage divider of the type shown in Fig. 4.34. Although simple, this component can generate a variety of problems. Of these, the first is that due to the value chosen for R_1. Unless this is infinitely high, it will attenuate the maximum signal voltage, (E_{max}), obtainable from the source, in the ratio

$$E_{max} = E_n \times R_1/(R_1 + Z_{source})$$

where Z_{source} is the output impedance of the driving circuit. This factor favours the use of a high value for R_1, to avoid loss of input signal.

However, the following amplifier stage may have specific input impedance requirements, and is unlikely to operate satisfactorily unless the output impedance of the gain control circuit is fairly low. This will vary according to the setting of the control, between zero and a value, at the maximum gain setting of the control, due to the parallel impedances of the source and gain control.

$$Z_{out} = R_1/(R_1 + Z_{source})$$

The output impedance at intermediate positions of the control varies as the effective source impedance and the impedance to the 0 V line is altered. However, in general, these factors would encourage the use of a low value for R_1.

An additional and common problem arises because the perceived volume level associated with a given sound pressure (power) level has a logarithmic characteristic. This means that the gain control potentiometer, R_1, must

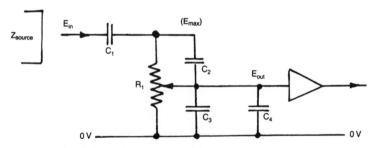

Fig. 4.34 *Standard gain control circuit.*

have a resistance value which has a logarithmic, rather than linear, relationship with the angular rotation of the potentiometer shaft.

POTENTIOMETER LAW

Since the most common types of control potentiometer employ a resistive composition material to form the potentiometer track, it is a difficult matter to ensure that the grading of conductivity within this material will follow an accurate logarithmic law.

On a single channel this error in the relationship between signal loudness and spindle rotation may be relatively unimportant. In a stereo system, having two ganged gain control spindles, intended to control the loudness of the two channels simultaneously, errors in following the required resistance law, existing between the two potentiometer sections, will cause a shift in the apparent location of the stereo image as the gain control is adjusted, and this can be very annoying.

In high-quality equipment, this problem is sometimes avoided by replacing R_1 by a precision resistor chain $(R_a–R_z)$, as shown in Fig. 4.35, in which the junctions between these resistors are connected to tapping points on a high-quality multi-position switch.

By this means, if a large enough number of switch tap positions is available, and this implies at least a 20-way switch to give a gentle gradation of sound level, a very close approximation to the required logarithmic law can be obtained, and two such channel controls could be ganged without unwanted errors in differential output level.

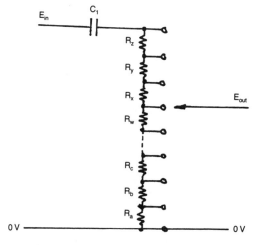

Fig. 4.35 *Improved gain control using multi-pole switch.*

A further practical problem, illustrated in Fig. 4.34, is associated with circuit capacitances. Firstly, it is essential to ensure that there is no standing DC potential across R_1 in normal operation, otherwise this will cause an unwanted noise in the operation of the control. This imposes the need for a protective input capacitor, C_1, and this will cause a loss of low frequency signal components, with a -3 dB LF turn-over point at the frequency at which the impedance of C_1 is equal to the sum of the source and gain control impedances. C_1 should therefore be of adequate value.

Additionally, there are the effects of the stray capacitances, C_2 and C_3, associated with the potentiometer construction, and the amplifier input and wiring capacitances, C_4. The effect of these is to modify the frequency response of the system, at the HF end, as a result of signal currents passing through these capacitances. The choice of a low value for R_1 is desirable to minimise this problem.

The use of the gain control to operate an on/off switch, which is fairly common in low-cost equipment, can lead to additional problems, especially with high resistance value gain control potentiometers, in respect of AC mains 'hum' pick up. It also leads to a more rapid rate of wear of the gain control, in that it is rotated at least twice whenever the equipment is used.

Tone controls

These exist in the various forms shown in Figs 4.36–4.40, respectively described as standard (bass and treble lift or cut), slope control, Clapham Junction, parametric and graphic equaliser types. The effect these will have on the frequency response of the equipment is shown in the drawings, and their purpose is to help remedy shortcomings in the source programme material, the receiver or transducer, or in the loudspeaker and listening room combination.

To the hi-fi purist, all such modifications to the input signal tend to be regarded with distaste, and are therefore omitted from some hi-fi equipment. However, they can be useful, and make valuable additions to the audio equipment, if used with care.

These are either of the passive type, of which a typical circuit layout is shown in Fig. 4.41, or are constructed as part of the negative feedback loop around a gain block, using the general design due to Baxandall. A typical circuit layout for this kind of design is shown in Fig. 4.42.

It is claimed that the passive layout has an advantage in quality over the active (feedback network) type of control, in that the passive network

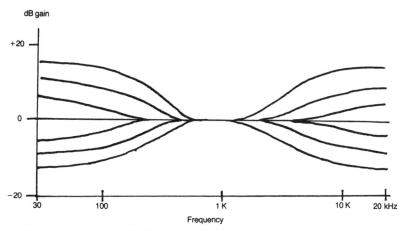

Fig. 4.36   *Bass and treble lift/cut tone control.*

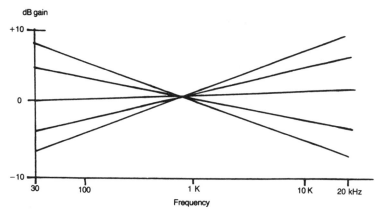

Fig. 4.37   *Slope control.*

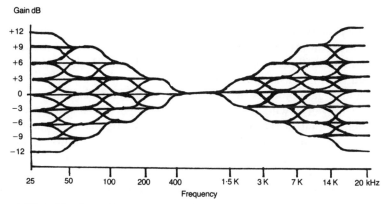

Fig. 4.38   *Clapham Junction type of tone control.*

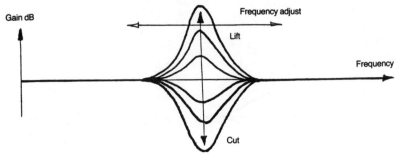

Fig. 4.39 *Parametric equaliser control.*

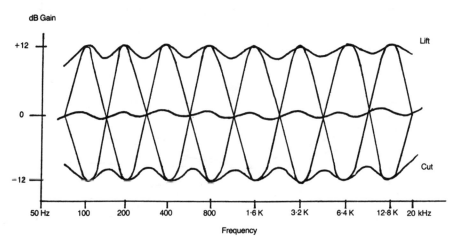

Fig. 4.40 *Graphic equaliser response characteristics.*

merely contains resistors and capacitors, and is therefore free from any possibility of introduced distortion, whereas the 'active' network requires an internal gain block, which is not automatically above suspicion.

In reality, however, any passive network must introduce an attenuation, in its flat response form, which is equal to the degree of boost sought at the maximum 'lift' position, and some external gain block must therefore be added to compensate for this gain loss.

This added gain block is just as prone to introduce distortion as that in an active network, with the added disadvantage that it must provide a gain equal to that of the flat-response network attenuation, whereas the active system gain block will typically have a gain of unity in the flat response mode, with a consequently lower distortion level.

As a final point, it should be remembered that any treble lift circuit will cause an increase in harmonic distortion, simply because it increases the gain at the frequencies associated with harmonics, in comparison with that at the frequency of the fundamental.

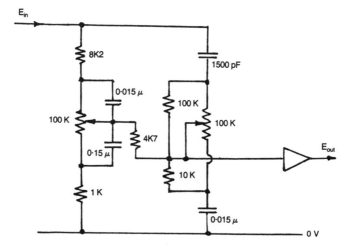

Fig. 4.41 *Circuit layout of passive tone control.*

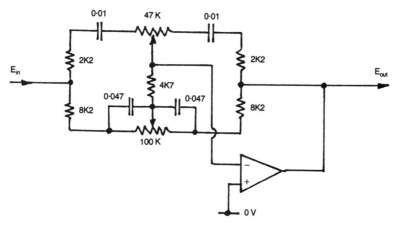

Fig. 4.42 *Negative feedback type tone control circuit.*

The verdict of the amplifier designers appears to be substantially in favour of the Baxandall system, in that this is the layout most commonly employed.

Both of these tone control systems – indeed this is true of all such circuitry – rely for their operation on the fact that the AC impedance of a capacitor will depend on the applied frequency, as defined by the equation:

$$Z_c = 1/(2\pi f_C),$$

or more accurately,

$$Z_c = 1/(2j\pi f_C),$$

where j is the square root of -1.

Commonly, in circuit calculations, the $2\pi f$ group of terms are lumped together and represented by the Greek symbol ω.

The purpose of the j term, which appears as a 'quadrature' element in the algebraic manipulations, is to permit the circuit calculations to take account of the 90° phase shift introduced by the capacitative element. (The same is also true of inductors within such a circuit, except that the phase shift will be in the opposite sense.) This is important in most circuits of this type.

The effect of the change in impedance of the capacitor on the output signal voltage from a simple RC network, of the kind shown in Figs 4.43(a) and 4.44(a), is shown in Figs 4.43(b) and 4.44(b). If a further

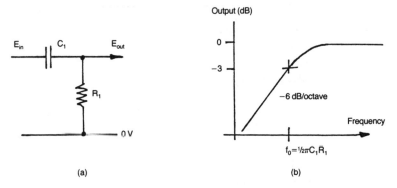

Fig. 4.43 *Layout and frequency response of simple bass-cut circuit (high-pass).*

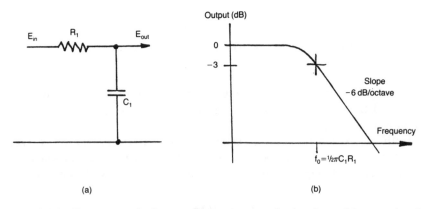

Fig. 4.44 *Layout and frequency response of simple treble-cut circuit (low-pass).*

resistor, R_2, is added to the networks the result is modified in the manner shown in Figs 4.45 and 4.46. This type of structure, elaborated by the use of variable resistors to control the amount of lift or fall of output as a function of frequency, is the basis of the passive tone control circuitry of Fig. 4.41.

If such networks are connected across an inverting gain block, as shown in Figs 4.47(a) and 4.48(a), the resultant frequency response will be shown in Figs 4.47(b) and 4.48(b), since the gain of such a negative feedback configuration will be

$$\text{Gain} = Z_a/Z_b$$

assuming that the open-loop gain of the gain block is sufficiently high. This is the design basis of the Baxandall type of tone control, and a flat frequency response results when the impedance of the input and output limbs of such a feedback arrangement remains in equality as the frequency is varied.

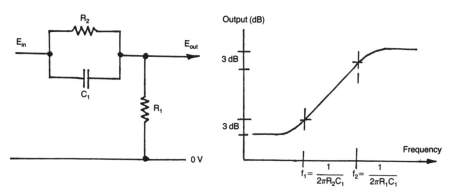

Fig. 4.45 *Modified bass-cut (high-pass) RC circuit.*

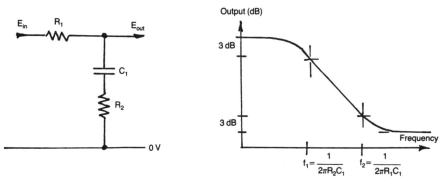

Fig. 4.46 *Modified treble-cut (low-pass) RC circuit.*

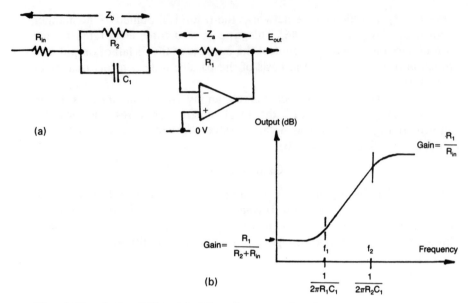

Fig. 4.47 *Active RC treble-lift or bass-cut circuit.*

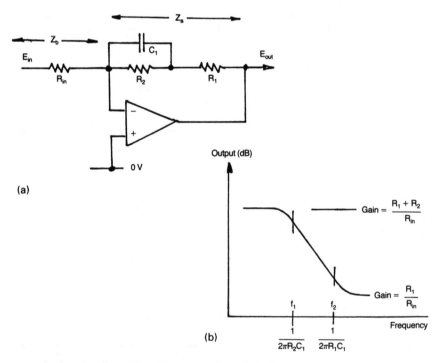

Fig. 4.48 *Active RC treble-cut or bass-lift circuit.*

SLOPE CONTROLS

This is the type of tone control employed by Quad in its type 44 pre-amplifier, and operates by altering the relative balance of the LF and HF components of the audio signal, with reference to some specified mid-point frequency, as is shown in Fig. 4.37. A typical circuit for this type of design is shown in Fig. 4.49.

The philosophical justification for this approach is that it is unusual for any commercially produced programme material to be significantly in error in its overall frequency characteristics, but the tonal preferences of the recording or broadcasting balance engineer may differ from those of the listener.

In such a case, he might consider that the signal, as presented, was somewhat over heavy, in respect of its bass, or alternatively, perhaps, that it was somewhat light or thin in tone, and an adjustment of the skew of the frequency response could correct this difference in tonal preference without significantly altering the signal in other respects.

THE CLAPHAM JUNCTION TYPE

This type of tone control, whose possible response curves are shown in Fig. 4.38, was introduced by the author to provide a more versatile type of tonal adjustment than that offered by the conventional standard systems, for remedying specific peaks or troughs in the frequency response, without the penalties associated with the graphic equaliser type of control, described below.

In the Clapham Junction type system, so named because of the similarity of the possible frequency response curves to that of railway lines, a group of push switches is arranged to allow one or more of a multiplicity of RC

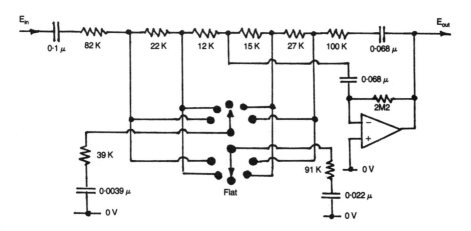

Fig. 4.49 *The Quad tilt control.*

networks to be introduced into the feedback loop of a negative feedback type tone control system, as shown in Fig. 4.50, to allow individual ±3 dB frequency adjustments to be made, over a range of possible frequencies.

By this means, it is possible by combining elements of frequency lift or cut to choose from a variety of possible frequency response curves, without losing the ability to attain a linear frequency response.

PARAMETRIC CONTROLS

This type of tone control, whose frequency response is shown in Fig. 4.39, has elements of similarity to both the standard bass/treble lift/cut systems, and the graphic equaliser arrangement, in that while there is a choice of lift or cut in the frequency response, the actual frequency at which this occurs may be adjusted, up or down, in order to attain an optimal system frequency response.

A typical circuit layout is shown in Fig. 4.51.

THE GRAPHIC EQUALISER SYSTEM

The aim of this type of arrangement is to compensate fully for the inevitable peaks and troughs in the frequency response of the audio system – including those due to deficiencies in the loudspeakers or the listening room acoustics – by permitting the individual adjustment of the channel gain, within any one of a group of eight single-octave segments of the frequency band, typically covering the range from 80 Hz to 20 kHz, though ten octave equalisers covering the whole audio range from 20 Hz to 20 kHz have been offered.

Because the ideal solution to this requirement – that of employing a group of parallel connected amplifiers, each of which is filtered so that it covers a single octave band of the frequency spectrum, whose individual gains could be separately adjusted – would be excessively expensive to implement, conventional practice is to make use of a series of LC tuned circuits, connected within a feedback control system, as shown in Fig. 4.52.

This gives the type of frequency response curve shown in Fig. 4.40. As can be seen, there is no position of lift or cut, or combination of control settings, which will permit a flat frequency response, because of the interaction, within the circuitry, between the adjacent octave segments of the pass-band.

While such types of tone control are undoubtedly useful, and can make significant improvements in the performance of otherwise unsatisfactory hi-fi systems, the inability to attain a flat frequency response when this is desired, even at the mid-position of the octave-band controls, has given such arrangements a very poor status in the eyes of the hi-fi fraternity. This unfavourable opinion has been reinforced by the less than optimal

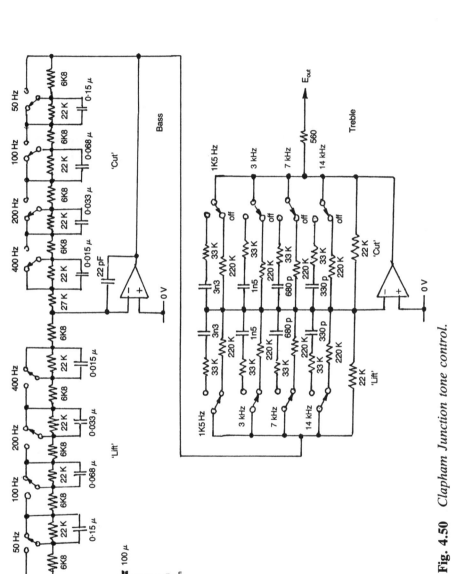

Fig. 4.50 *Clapham Junction tone control.*

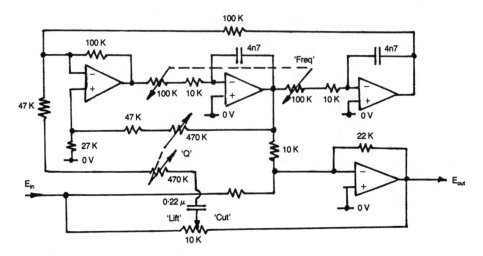

Fig. 4.51   *Parametric equaliser circuit.*

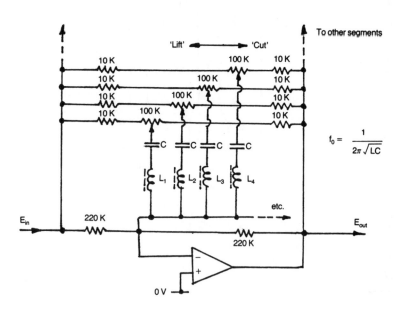

Fig. 4.52   *Circuit layout for graphic equaliser (four sections only shown).*

performance offered by inexpensive, add-on, units whose engineering standards have reflected their low purchase price.

Channel balance controls

These are provided in any stereo system, to equalise the gain in the left- and right-hand channels, to obtain a desired balance in the sound image. (In a quadraphonic system, four such channel gain controls will ideally be provided.) In general, there are only two options available for this purpose: those balance controls which allow one or other of the two channels to be reduced to zero output level, and those systems, usually based on differential adjustment of the amount of negative feedback across controlled stages, in which the relative adjustment of the gain, in one channel with reference to the other, may only be about 10 dB.

This is adequate for all balance correction purposes, but does not allow the complete extinction of either channel.

The first type of balance control is merely a gain control, of the type shown in Fig. 4.34. A negative feedback type of control is shown in Fig. 4.53.

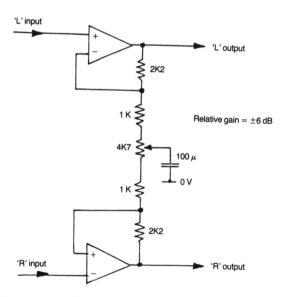

Fig. 4.53 *Negative feedback type channel balance control.*

Channel separation controls

While the closest reproduction, within the environment of the listener, of the sound stage of the original performance will be given by a certain specific degree of separation between the signals within the 'L' and 'R' channels, it is found that shortcomings in the design of the reproducing and amplifying equipment tend universally to lessen the degree of channel separation, rather than the reverse.

Some degree of enhancement of channel separation is therefore often of great value, and electronic circuits for this purpose are available, such as that, due to the author, shown in Fig. 4.54.

There are also occasions when a deliberate reduction in the channel separation is of advantage, as, for example, in lessening 'rumble' effects due to the vertical motion of a poorly engineered record turntable, or in lessening the hiss component of a stereo FM broadcast. While this is also provided by the circuit of Fig. 4.54, a much less elaborate arrangement, as shown in Fig. 4.55, will suffice for this purpose.

A further, and interesting, approach is that offered by Blumlein, who found that an increase or reduction in the channel separation of a stereo signal was given by adjusting the relative magnitudes of the 'L+R' and 'L−R' signals in a stereo matrix, before these were added or subtracted to give the '2L' and '2R' components.

An electronic circuit for this purpose is shown in Fig. 4.56.

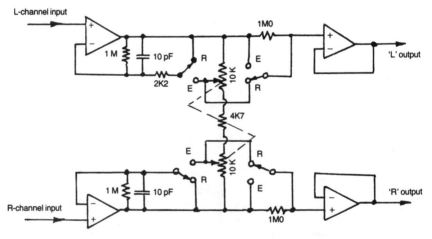

Fig. 4.54 *Circuit for producing enhanced or reduced stereo channel separation.*

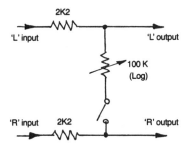

Fig. 4.55 *Simple stereo channel blend control.*

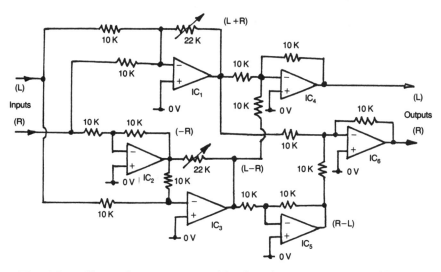

Fig. 4.56 *Channel separation or blending by using matrix addition or subtraction.*

Filters

While various kinds of filter circuit play a very large part in the studio equipment employed to generate the programme material, both as radio broadcasts and as recordings on disc or tape, the only types of filter normally offered to the user are those designed to attenuate very low frequencies, below, say, 50 Hz and generally described as 'rumble' filters, or those operating in the region above a few kHz, and generally described as 'scratch' or 'whistle' filters.

Three such filter circuits are shown in Figs 4.57(a), (b), and (c). Of these the first two are fixed frequency active filter configurations employing

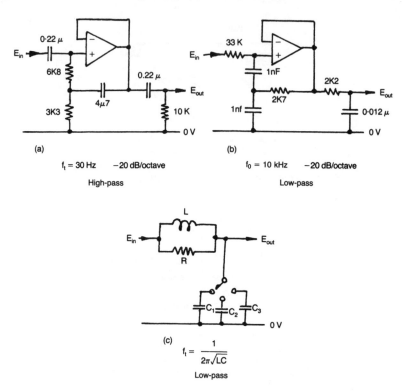

Fig. 4.57 *Steep-cut filter circuits.*

a bootstrap type circuit, for use respectively in high-pass (rumble) and low-pass (hiss) applications, and the third is an inductor-capacitor passive circuit layout, which allows adjustment of the HF turn-over frequency by variation of the capacitor value.

Such frequency adjustments are, of course, also possible with active filter systems, but require the simultaneous switching of a larger number of components. For such filters to be effective in their intended application, the slope of the response curve, as defined as the change in the rate of attenuation as a function of frequency, is normally chosen to be high – at least 20 dB/octave – as shown in Fig. 4.58, and, in the case of the filters operating in the treble region, a choice of operating frequencies is often required, as is also, occasionally, the possibility of altering the attenuation rate.

This is of importance, since rates of attenuation in excess of 6 dB/octave lead to some degree of coloration of the reproduced sound, and the greater the attenuation rate, the more noticeable this coloration becomes. This problem becomes less important as the turn-over frequency approaches the limits of the range of human hearing, but very steep rates of

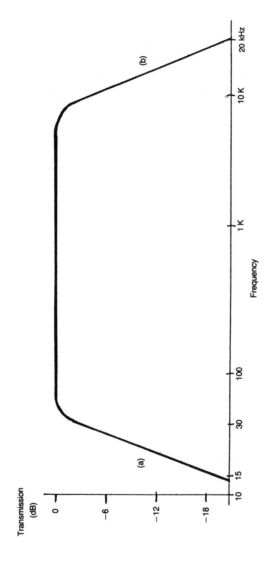

Fig. 4.58 *Characteristics of circuits of Figs 4.57(a) and (b).*

attenuation produce distortions in transient waveforms whose major frequency components are much lower than notional cut-off frequency.

It is, perhaps, significant in this context that recent improvements in compact disc players have all been concerned with an increase in the sampling rate, from 44.1 kHz to 88.2 kHz or 176.4 kHz, to allow more gentle filter attenuation rates beyond the 20 kHz audio pass-band, than that provided by the original 21 kHz 'brick wall' filter.

The opinion of the audiophiles seems to be unanimous that such CD players, in which the recorded signal is two- or four-times 'oversampled', which allows much more gentle 'anti-aliasing' filter slopes, have a much preferable HF response and also have a more natural, and less prominent, high-frequency characteristic, than that associated with some earlier designs.

CHAPTER 5

Power output stages

In principle, the function of an audio power amplifier is a simple one: to convert a low-power input voltage signal from a pre-amplifier or other signal source into a higher power signal capable of driving a loudspeaker or other output load, and to perform this amplification with the least possible alteration of the input waveform or to the gain/frequency or phase/frequency relationships of the input signal.

VALVE AMPLIFIER DESIGNS

In the earlier days of audio amplifiers, this function was performed by a two- or three-valve circuit design employing a single output triode or pentode, transformer coupled to the load, and offering typical harmonic distortion levels, at full power – which would be, perhaps, in the range 2–8 W – of the order of 5–10% THD, at 1 kHz, and amplifier output stages of this kind formed the mainstay of radio receivers and radiograms in the period preceding and immediately following the Second World War.

With the improvement in the quality of gramophone recordings and radio broadcasts following the end of the war, and the development of the new 'beam tetrode' as a replacement for the output pentode valve, the simple 'single-ended' triode or pentode output stage was replaced by push–pull output stages. These used beam tetrodes either triode-connected, as in the celebrated Williamson design, or with a more complex output transformer design, as in the so-called Ultra-linear layout, or in the Quad design, shown in Figs 5.1(a), (b) and (c).

The beam tetrode valve construction, largely the work of the Marconi–Osram valve company of London, offered a considerable improvement in the distortion characteristics of the output pentode, while retaining the greater efficiency of that valve in comparison with the output triode. This advantage in efficiency was largely retained when the second grid was connected to the anode so that it operated as a triode.

This electrode interconnection was adopted in the Williamson design, shown in Fig. 5.2, in which a substantial amount of overall negative feedback was employed, to give a harmonic distortion figure of better than 0.1%, at the rated 15 W output power.

189

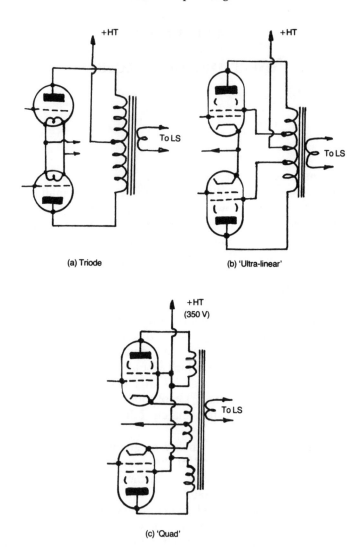

(a) Triode

(b) 'Ultra-linear'

(c) 'Quad'

Fig. 5.1 *Push–pull valve amplifier output stages.*

In general, the principle quality determining factor of such an audio amplifier was the output transformer, which coupled the relatively high output impedance of the output valves to the low impedance of the loudspeaker load, and good performance demanded a carefully designed and made component for this position. Nevertheless, such valve amplifier designs did give an excellent audio performance, and even attract a nostalgic following to this day.

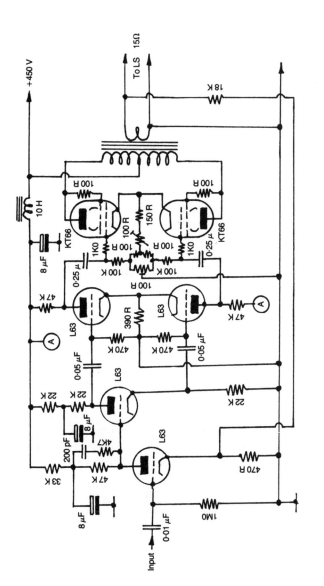

Fig. 5.2 *The Williamson amplifier.*

EARLY TRANSISTOR CIRCUITS

With the advent of semiconductors with significant power handling capacity, and the growing confidence of designers in their use, transistor audio amplifier designs began to emerge, largely based on the 'quasi-complementary' output transistor configuration designed by HC Lin, which is shown in Fig. 5.3. This allowed the construction of a push–pull power output stage in which only PNP power transistors were employed, these being the only kind then available. It was intended basically for use with the earlier diffused junction Germanium transistors, with which it worked adequately well.

However, Germanium power transistors at that time had both a limited power capability and a questionable reliability, due to their intolerance of junction temperatures much in excess of 100° C. The availability of silicon power transistors, which did not suffer from this problem to such a marked degree, prompted the injudicious appearance of a crop of solid-state audio amplifiers from manufacturers whose valve operated designs had enjoyed an erstwhile high reputation in this field.

LISTENER FATIGUE AND CROSSOVER DISTORTION

Fairly rapidly after these new semiconductor-based audio amplifiers were introduced, murmurings of discontent began to be heard from their users, who claimed, quite justifiably, that these amplifiers did not give the same warmth and fullness of tone as the previous valve designs, and the term 'listener fatigue' was coined to describe the user's reaction.

The problem, in this case, was a simple one. In all previous audio amplifier experience, it had been found to be sufficient to measure the distortion and other performance characteristics at full output power, since it could reasonably be assumed that the performance of the amplifier would improve as the power level was reduced. This was not true for these early transistor audio amplifiers, in which the performance at full power was probably the best it would ever give.

This circumstance arose because, although the power transistors were capable of increased thermal dissipation, it was still inadequate to allow the output devices to be operated in the 'Class A' mode, always employed in conventional valve operated power output stages, in which the quiescent operating current, under zero output power conditions, was the same as that at maximum output.

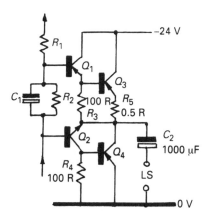

Fig. 5.3 *Quasi-complementary push–pull transistor output stage from HC Lin.*

Crossover problems

It was normal practice, therefore, with transistor designs, to bias the output devices into 'Class B' or 'Class AB', in which the quiescent current was either zero, or substantially less than that at full output power, in order to increase the operating efficiency, and lessen the thermal dissipation of the amplifier.

This led to the type of defect known as 'crossover distortion', which is an inherent problem in any class 'B' or 'AB' output stage, and is conspicuous in silicon junction transistors, because of the abrupt turn-on of output current with increasing base voltage, of the form shown in Fig. 5.4.

In a push–pull output stage employing such devices, the transfer characteristics would be as shown in Fig. 5.5(b), rather than the ideal straight line transfer of Fig. 5.5(a). This problem is worsened anyway, in the quasi-complementary Lin design, because the slopes of the upper and lower transfer characteristics are different, as shown in Fig. 5.5(c).

The resulting kink in the transfer characteristic shown in Fig. 5.5(d) lessens or removes the gain of the amplifier for low-level signals which cause only a limited excursion on either side of the zero voltage axis, and leads to a 'thin' sound from the amplifier.

It also leads to a distortion characteristic which is substantially worse at low power levels − at which much listening actually takes place − than at full power output. Moreover, the type of distortion products generated by crossover distortion comprise the dissonant seventh, ninth, eleventh and thirteenth harmonics, which are much more objectionable, audibly, than the second and third harmonics associated with the previous valve amplifier designs.

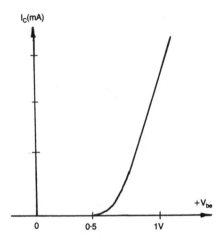

Fig. 5.4 *Turn-on characteristics of silicon junction transistor.*

Additionally, the absence of the output transformer from such transistor amplifier designs allowed the designers to employ much greater amounts of negative feedback (NFB) from the output to the input of the circuit, to attempt to remedy the non-linear transfer characteristics of the system. This increased level of NFB impaired both the load stability and the transient response of the amplifier, which led to an 'edgy' sound. High levels of NFB also led to a more precise and rigid overload clipping level. Such a proneness to 'hard' clipping also impaired the sound quality of the unit.

This failure on the part of the amplifier manufacturers to test their designs adequately before offering them for sale in what must be seen in retrospect as a headlong rush to offer new technology is to be deplored for several reasons.

Of these, the first is that it saddled many unsuspecting members of the public with equipment which was unsatisfactory, or indeed unpleasant, in use. The second is that it tended to undermine the confidence of the lay user in the value of specifications – simply because the right tests were not made, and the correct parameters were left unspecified. Finally, it allowed a 'foot in the door' for those whose training was not in the field of engineering, and who believed that technology was too important to be left to the technologists.

The growth of this latter belief, with its emphasis on subjective assessments made by self-appointed pundits, has exerted a confusing influence on the whole field of audio engineering, it has led to irrationally based choices in consumer purchases and preferences, so that many of the

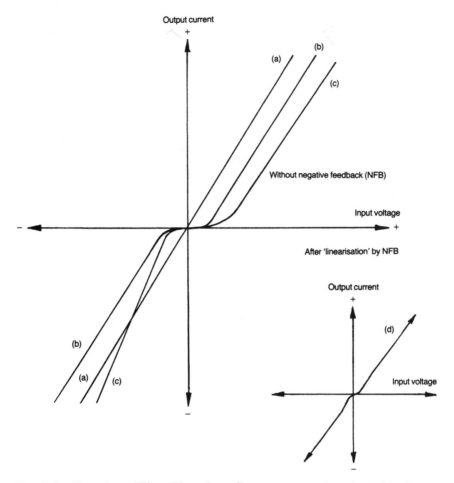

Fig. 5.5 *Transistor 'Class B' push–pull output stage characteristics. (Line 'a' is a straight line for comparison.)*

developments in audio engineering over the past two decades, which have been substantial, have occurred in spite of, and frequently in the face of, well-publicised and hostile opposition from these critics.

IMPROVED TRANSISTOR AMPLIFIER DESIGNS

The two major developments which occurred in transistor output circuit technology, which allowed the design of transistor power amplifiers which had a low inherent level of crossover distortion, were the introduction in the mid 1960s of fully complementary (NPN and PNP) output power transistors by Motorola Semiconductors Inc, and the invention by IM Shaw (*Wireless World*, June 1969), subsequently refined by P. J. Baxandall (*Wireless World*, September 1969) of a simple circuit modification to the basic quasi-complementary output push–pull pair to increase its symmetry.

These layouts are shown in Figs 5.6(a) and (b). The modified version shown in Fig. 5.6(c) is that used by the author in his *Hi-Fi News* 75 W power amplifier design, for which the complete circuit is given in Fig. 5.7.

In many ways the Shaw/Baxandall quasi-complementary output transistor circuit is preferable to the use of fully complementary output transistors, since, to quote J. Vereker of *Naim Audio*, 'NPN and PNP power transistors are only really as equivalent as a man and a woman of the same weight and height' – the problem being that the different distribution of the N- and P-doped layers leads to significant differences in the HF performance of the devices. Thus, although the circuit may have a good symmetry at low frequencies, it becomes progressively less symmetrical with increasing operating frequency.

With modern transistor types, having a higher current gain transition frequency (that frequency at which the current gain decreases to unity), the HF symmetry of fully complementary output stages is improved, but it is still less good than desired, so the relative frequency/phase characteristics of each half of the driver circuit may need to be adjusted to obtain optimum performance.

POWER MOSFETs

These transistor types, whose operation is based on the mobility of electrostatically induced charge layers (electrons or 'holes') through a very narrow layer of undoped or carrier-depleted silicon, have a greatly superior performance to that of bipolar junction transistors, both in respect of maximum operating frequency and linearity, and allow considerable improvements in power amplifier performance for any given degree of circuit complexity.

Two forms of construction are currently employed for power MOSFETs, the vertical or 'V' MOSFET, which can employ either a 'V' or a 'U' groove formation, of the types shown in Figs 5.8(a) and (b); though, of these two, the latter is preferred because of the lower electrostatic stress levels associated with its flat-bottomed groove formation; or the 'D' MOSFET shown in Fig. 5.8(c).

These devices are typically of the 'enhancement' type, in which no current flows at zero gate/source voltage, but which begin to conduct, progressively, as the forward gate voltage is increased. Once the

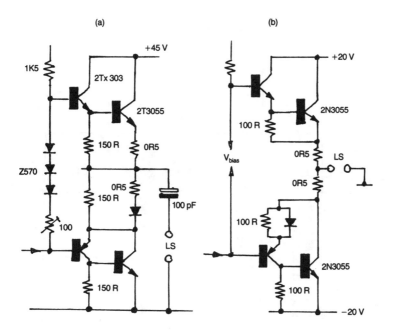

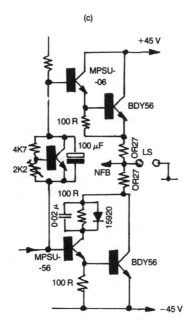

Fig. 5.6 *Improved push–pull transistor output stages.*

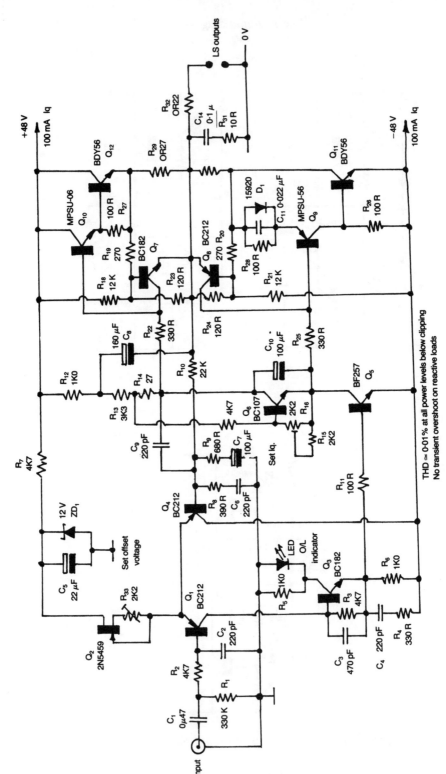

Fig. 5.7 *JLH Hi-Fi News 75 W audio amplifier.*

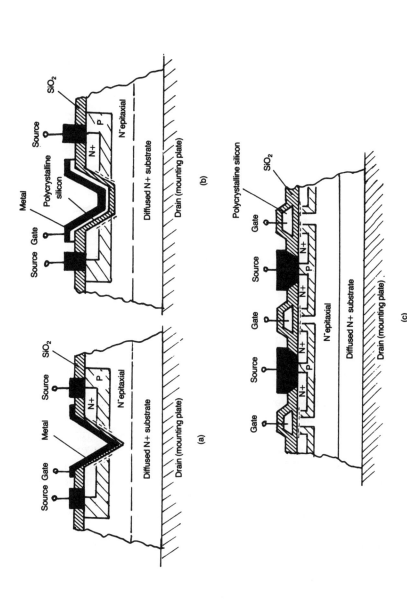

Fig. 5.8 Power MOSFET structure. (a) V-MOSFET, (b) U-MOSFET, (c) D-MOSFET.

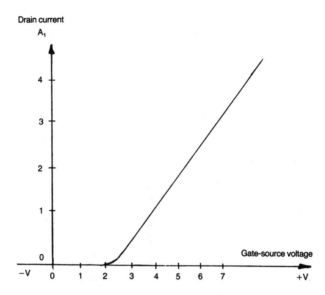

Fig. 5.9   *Power MOSFET operating characteristics.*

conduction region has been reached, the relationship between gate voltage and drain current is very linear, as shown in Fig. 5.9.

By comparison with 'bipolar' junction power transistors, of conventional types, the MOSFET, which can be made in both N channel and P
channel types to allow symmetrical circuit layout – though there is a
more limited choice of P channel devices – does not suffer from stored
charge effects which limit the 'turn−off' speed of bipolar transistors.

The greater speed and lesser internal phase shifts of power amplifiers
based on power MOSFETs allow greater freedom in the design of
overall NFB layouts. This in turn, gives superior performance in the
middle to upper frequency range. Greater care is needed, in the circuit
design, however to avoid high frequency parasitic oscillation, which can
cause rapid failure of the devices.

The principal technical problem in the evolution of power transistors
derived from these charge operated devices, as distinct from the existing
small signal MOSFETs, lies in obtaining an adequate capacity for current
flow through the device. This problem is solved in two ways; by making
the conduction path through which the current must flow as short as
possible, and by fabricating a multiplicity of conducting elements on the
surface of the silicon slice, which can be connected in parallel to lower the
conducting resistance of the device.

V and U types

In the 'V' or 'U' groove MOSFET, taking the case of the 'N' channel devices shown in Fig. 5.8, the required narrow conduction channel, in which electrostatically formed negative charges (electrons) may be induced in the depleted 'P' type layer, is obtained by etching a channel through a previously diffused, or epitaxially grown, pair of differently doped layers. With modern technology the effective thickness of such layers can be controlled with great precision.

In the case of the D-MOS device of Fig. 5.8(c), the required narrow channel length is achieved by the very precise geometry of the diffusion masks used to position the doped regions. It is customary in the D-MOS devices to use a polycrystalline silicon conducting layer, rather than aluminum, to provide the gate electrode, since this offers a lower likelihood of contamination of the thin gate insulating layer.

In all of these MOS devices, the method of operation is that a voltage applied to the gate electrode will cause a charge to be induced in the semiconductor layer immediately below it, which, since this layer of charge is mobile, will cause the device to conduct.

Both for the 'V/U' MOS and the 'D' MOS devices, the voltage breakdown threshold of the gate insulating layer, which must be thin if the device is to operate at all, will be in the range 15–35 V. It can be seen that part of this gate insulating layer lies between the gate electrode and the relatively high voltage 'drain' electrode. Avoidance of gate/drain voltage breakdown depends therefore on there being an adequate voltage drop across the N-doped intervening drain region. This in turn depends on the total current flow through the device.

This has the effect that the actual gate/drain breakdown voltage is output current-dependent, and that some protective circuitry may need to be used, as in the case of bipolar output transistors, if damage is to be avoided.

OUTPUT TRANSISTOR PROTECTION

An inconvenient characteristic of all junction transistors is that the forward voltage of the P–N junction decreases as its temperature is increased. This leads to the problem that if the current through the device is high enough to cause significant heating of the base-emitter region, the forward voltage drop will decrease. If, due to fortuitous variations in the thickness of this layer or in its doping level, some regions of this junction area heat up more than others, then the forward voltage drop of these areas will be less, and current will tend to be funnelled through these areas causing yet further heating.

This causes the problem known as 'thermal runaway' and the phenomenon of secondary breakdown, if certain products of operating current and voltage are exceeded. The permitted regions of operation for any particular bipolar transistor type will be specified by the manufacturers in a 'safe operating area' (SOA) curve, of the type shown in Fig. 5.10.

The circuit designer must take care that these safe operating conditions are not exceeded in use, for example by the inadvertent operation of the amplifier into a short-circuited output load. Provided that the SOA limits are not greatly exceeded, a simple fuse on the output circuit will probably be adequate, but a more effective type of protection is that given by the clamp transistors, Q_7 and Q_8 in the power amplifier design shown in Fig. 5.7.

In this circuit arrangement, designed by A. R. Bailey, the clamp transistors monitor simultaneously the voltage present across the output transistors, by means, for example, of R_{18} and R_{23}, and also the output current, in the case of the upper output transistor, by monitoring the voltage developed across R_{29}. If the combination of these two voltage contributions exceeds the 0.55 V turn-on voltage of the clamp transistor, (Q_7), it will conduct and shunt the input signal applied to the first transistor of the output pair (Q_{10}) and avoid output transistor damage.

In the case of power MOSFETs, a simple zener diode, connected to limit the maximum forward voltage which can be applied to the output device, may be quite adequate. An example of this type of output stage

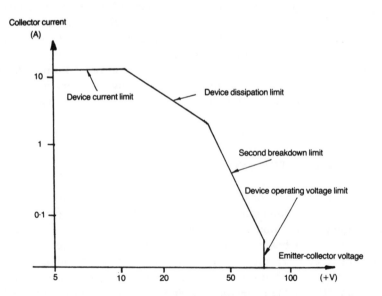

Fig. 5.10 *Typical safe operating area (SOA) curve for junction power transistor.*

protection is shown in the circuit for the output stages of a 45 W power MOSFET amplifier designed by the author, shown in Fig. 5.11.

POWER OUTPUT AND POWER DISSIPATION

One of the most significant changes in audio technology during the past 30 years has been in the design of loudspeaker units, in which the electro-acoustic efficiency, in terms of the output sound level for a given electrical power input, has been progressively traded off against flatness of frequency response and reduced coloration.

This is particularly notable in respect of the low-frequency extension of the LS response, in closed box 'infinite baffle' systems. Among other changes here, the use of more massive bass driver diaphragms to lower the fundamental resonant frequency of the system − below which a −12 dB/octave fall-off in sound output will occur − requires that more input power is required to produce the same diaphragm acceleration.

On the credit side, the power handling capacity of modern LS systems has also been increased, so that equivalent or greater sound output levels are still obtainable, but at the cost of much greater input power levels.

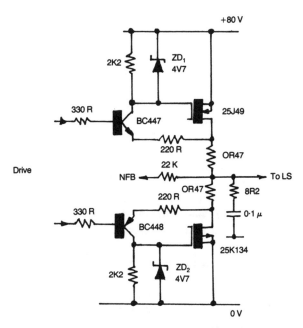

Fig. 5.11   *Zener diode protection for power MOSFET output stage.*

Power levels

As a specific example of the power levels which may be needed for realistic reproduction of live sounds using typical modern LS units, recent measurements made in a recording studio showed that the peak output power required from the amplifier to match the peak sound levels produced by a grand piano played by a professional musician in the same room, was in excess of 300 W per channel. This test was made using a pair of high-quality monitor speakers, of a kind also widely used in domestic hi-fi set-ups, whose overall efficiency was typical of modern designs.

No conventional valve amplifier, operating in 'Class A', could meet this power output requirement, in a stereo system, without the penalties incurred in terms of size, cost, and heat output being intolerable. In contrast the solid-state class 'AB' power amplifier actually used in this test fitted comfortably into a 7 in. high slot in a standard 19 in. rack, and ran cool in use.

This reduction in LS efficiency has led to compensatory increases in power amplifier capability, so that the typical output power of a contemporary domestic audio amplifier will be in the range 50–100 W, measured usually with a 4 or 8 ohm resistive load, with 25 or 30 W units, which would at one time have been thought to be very high power units, now being restricted to budget priced 'mini' systems.

Design requirements

The design requirements for such power output levels may be seen by considering the case of a 100 W power amplifier using junction transistors in an output stage circuit of the form shown in Fig. 5.12, the current demand is given by the formula.

$$I = \sqrt{P/R}$$

where $P = 100$, and $R = 8$. This gives an output current value of 3.53 A (RMS), equivalent to a peak current of 5 A for each half cycle of a sinusoidal output signal. At this current the required base-emitter potential for Q_3 will be about 3 V, and allowing for the collector-emitter voltage drop of Q_1, the expected emitter-collector voltage in Q_3 will be of the order of 5 V.

In an 8 ohm load, a peak current of 5 A will lead to a peak voltage, in each half cycle, of 40 V. Adding the 1.1 V drop across R_3, it can be seen that the minimum supply voltage for the positive line supply must be at least 46.1 V. Since the circuit is symmetrical, the same calculations will apply to the potential of the negative supply line. However, in any practical design, the supply voltages chosen will be somewhat greater

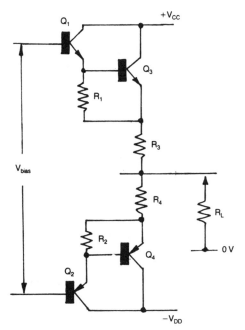

Fig. 5.12   *Typical transistor output stage.*

than the calculated minimum, so line voltages of ±50 V would probably be used.

For a true 'Class A' system, in which both of the output transistors remained in conduction for the whole output cycle, an operating current of at least 5 A, DC, would be required, leading to an output stage dissipation, for each channel, of 500 W.

In the case of a typical 'Class AB' output stage, with a quiescent current of 100 mA, a much more modest no-signal dissipation of 10 W would be required. At full power, the device dissipation may be calculated from the consideration that the RMS current into the load, in each half cycle, will be 3.53 A giving an RMS voltage swing of 28.24 V. The mean voltage across each device during the half cycle of its conduction would therefore be 50 minus 28.24, and the mean dissipation, for each half of the power output stage would be 76.8 W.

The worst case dissipation for such output stages, depending on supply line voltage margins, will be at about half power, where the dissipation in each half of the output stage could rise to 80–85 W, and the heat sinking arrangements provided must be adequate to meet this need. It is still, however, only a third of that required by a 'Class A' system, and even then only required on power peaks, which, with typical programme material, occur infrequently so that the average output stage heat dissipation, even under conditions where the amplifier is used at near its rated maximum power output, may well only be 20–30 W.

GENERAL DESIGN CONSIDERATIONS

Some of the techniques used for obtaining high gain and good linearity from a voltage amplifier stage, using semiconductors, were discussed in Chapter 4. However, in the case of low-power voltage amplifier stages, all of the gain elements will be operated in 'Class A', so that there will not be a need to employ large amounts of overall negative feedback (NFB) to assist in linearising the operating characteristics of the stage.

In power amplifier circuits, on the other hand, the output devices will almost certainly be operated in a non-linear part of their characteristics, where a higher level of NFB will be needed to reduce the associated distortion components to an acceptable level.

Other approaches have been adopted in the case of power amplifier circuits, such as the feed-forward of the error signal, or the division of the output stage into low power 'Class A' and high power 'Class B' sections. These techniques will be discussed later. Nevertheless, even in these more sophisticated designs, overall NFB will still be employed, and the problems inherent in its use must be solved if a satisfactory performance is to be obtained.

Of these problems, the first and most immediate is that of feedback loop stability. The Nyquist criterion for stability in any system employing negative feedback is that, allowing for any attenuation in the feedback path, the loop gain of the system must be less than unity at any frequency at which the loop phase shift reaches 180°.

Bode Plot

This requirement is most easily shown in the 'Bode Plot' of gain versus frequency illustrated in Fig. 5.13. In the case of a direct-coupled amplifier circuit, and most modern transistor systems will be of this type, the low-frequency phase lead is unlikely to exceed 90°, even at very low frequencies, but at the higher frequency end of the spectrum the phase lag of the output voltage in relation to the input signal will certainly be greater than 180° at the critical unity gain frequency, in any amplifier employing more than two operational stages, unless some remedial action is taken.

In the case of the simple power amplifier circuit shown schematically in Fig. 5.14, the input long-tailed pair, Q_1 and Q_2, with its associated constant current source, CC_1, and the second stage, 'Class A' amplier, Q_3

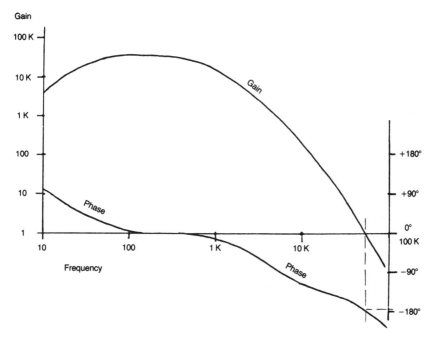

Fig. 5.13 *Bode Plot relating gain and phase to frequency in notional marginally stable amplifier.*

with its constant current load, CC_2, form the two voltage gain stages, with Q_4 and Q_5 acting as emitter-followers to transform the circuit output impedance down to a level which is low enough to drive a loudspeaker (Z_1).

In any amplifier having an output power in excess of a few watts, the output devices, Q_4 and Q_5, will almost certainly be elaborated to one of the forms shown in Figs 5.3 or 5.6, and, without some corrective action, the overall phase shift of the amplifier will certainly be of the order of 270°, at the critical unity gain point. This means that if overall negative feedback were to be employed, via R_5 and R_4, the circuit would oscillate vigorously at some lower, 180° phase-shift frequency.

SLEW-RATE LIMITING AND TID

Until a few years ago, the most common way of stabilising a negative feedback amplifier of this type was by means of a small capacitor, C_4, connected across the second amplifier transistor, Q_3. This particularly appealed to the circuit designers because it allowed the circuit to be

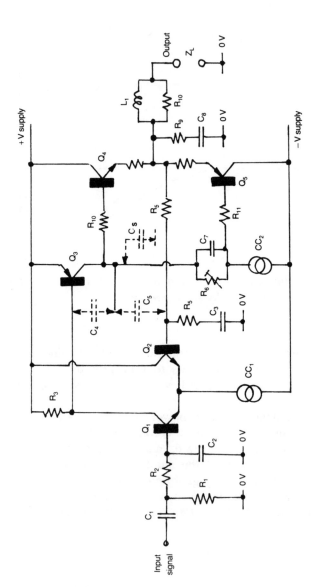

Fig. 5.14 *Basic layout of simple audio amplifier.*

stabilised while retaining a high level of closed-loop gain up to a high operating frequency, and this facilitated the attainment of low overall harmonic distortion figures at the upper end of the audio frequency band.

Unfortunately, the use of an HF loop stabilisation capacitor in this position gave rise to the problem of slew-rate limiting, because there is only a finite rate at which such a capacitor could charge through the constant current source, CC_2, or discharge through the current available at the collector of Q_1.

Slew limiting

This leads to the type of phenomenon shown in Fig. 5.15, in which, if a continuous signal occurs at the same time as one which leads to a sudden step in the mean voltage level – which could happen readily in certain types of programme material – then the continuous signal will either be mutilated or totally lost during the period during which the amplifier output voltage traverses, (slews), under slew-rate limited conditions, between output levels '1' and '2'.

This type of problem had been well known among amplifier designers for many years, and the deleterious effect upon the sound of the amplifier had also been appreciated by those who were concerned about this point. However, the commercial pressures upon designers to offer circuits which met the reviewers magic specification of 'less that 0.02% THD from 20 Hz to 20 kHz', led to many of the less careful, or more cynical, to use the slew-rate limiting system of HF stabilisation, regardless of the final sound quality.

Public attention was drawn to this phenomenon by M. Otala, in a paper in September 1970 (*IEE Trans.* AU–18, No 3), and the subsequent discussions and correspondence on this point were valuable in convincing the reviewers that low levels of harmonic distortion, on their own, were an insufficient guarantee of audio quality.

HF stabilisation

The alternative method of HF stabilisation which could be employed in the circuit of Fig. 5.14, and which does not lead to the same problems of slew-rate limitation – in that it causes the HF gain to decrease throughout the amplifier as a whole, rather than leaving the input stage operating without feedback at all during the slew-rate limited period – is by the use of capacitor between Q_3 collector, and Q_2 base, as indicated by C_5, and as used in the author's 15–20 W 'class AB' design (*Wireless World*, July 1970).

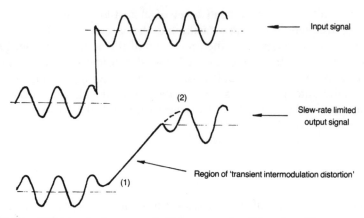

Input signal

(2)

Slew-rate limited
output signal

Region of 'transient intermodulation distortion'

(1)

Fig. 5.15 *Effect of slew-rate limiting on amplifier handling combined signals.*

Unfortunately, although this type of stabilisation leads to designs with pleasing sound quality, it does not lead to the same degree of distortion reduction at the HF end of the spectrum, unless a higher level of loop gain is employed. The 75 W design shown in Fig. 5.7, which achieves a typical distortion performance of less than 0.015% over the whole audio spectrum, is a good example of this latter approach.

A useful analysis of the problem of slew-rate induced distortion, in non-mathematical terms, was given by W. G. Jung, in *Hi-Fi News* (Nov. 1977), with some additional remedial circuit possibilities shown by the author in the same journal in January 1978.

Zobel network

Other components used for correcting the phase characteristics of the feedback loop are the so-called Zobel network (C_8, R_9) connected across the output, which will help avoid instability with an open-circuit load, and the inductor/resistor network (L_1, R_{10}) in series with the load, which is essential in some designs to avoid instability on a capacitative load.

Since L_1 is a very small inductor, typically having a value of 4.7 microhenries, its acoustic effects within the 20–20 kHz band will usually be inaudible. Its presence will, however, spoil the shape of any square-wave test waveform, as displayed on an oscilloscope, especially with a capacitative simulated LS load.

Bearing in mind that all of these slew-rate and transient waveform defects are worse at higher rates of change of the input signal waveform, it is prudent to include an input integrating network, such as R_2/C_2 in

Fig. 5.14, to limit the possible rate of change of the input signal, and such an input network is common on all contemporary designs. Since the ear is seldom sensitive beyond 20 kHz, except in young children, and most adults' hearing fails below this frequency, there seems little point in trying to exceed this HF bandwidth, except for the purpose of producing an impressive paper specification.

A further significant cause of slew-rate limitation is that of the stray capacitances (C_s) associated with the collector circuitry of Q_3, in Fig. 5.14, and the base input capacitance of Q_4/Q_5. This stray capacitance can be charged rapidly through Q_3, but can only be discharged again at a rate determined by the output current provided by the current source CC_2. This current must therefore be as large as thermal considerations will allow.

The possible avoidance of this source of slew-rate limitation by fully symmetrical amplifier systems, as advocated by Hafler, has attracted many designers.

Among the other defects associated with the power output stage, is that of 'hang up' following clipping, of the form shown graphically in Fig. 5.16. This has the effect of prolonging, and thereby making more audible, the effects of clipping. It can usually be avoided by the inclusion of a pair of drive current limiting resistors in the input circuitry of the output transistors, as shown by R_{10} and R_{11} in Fig. 5.14.

ADVANCED AMPLIFIER DESIGNS

In normal amplifier systems, based on a straightforward application of negative feedback, the effective circuit can be considered as comprising four parts:

- the output power stages, which will normally be simple or compound emitter followers, using power MOSFETs or bipolar transistors, using one of the forms shown in Figs 5.17(a)–(f);
- the voltage amplifying circuitry;
- the power supply system; and
- the protection circuitry needed to prevent damage to the output devices or the LS load.

The choice made in respect of any one of these will influence the requirements for the others.

Power supply systems

It has, for example, been common practice in the past to use a simple mains transformer, rectifier and reservoir capacitor combination of the

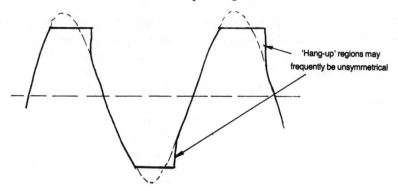

Fig. 5.16   *Effect of 'hang-up' following clipping.*

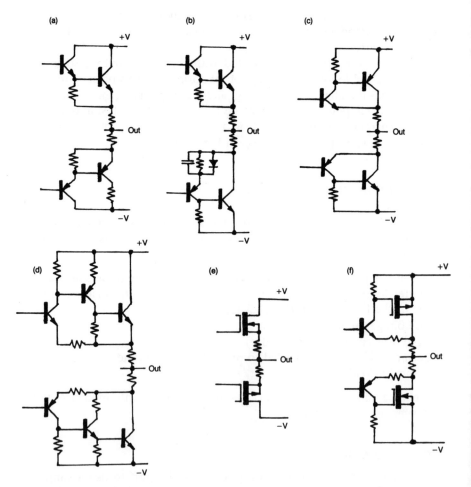

Fig. 5.17   *Output transistor circuits.*

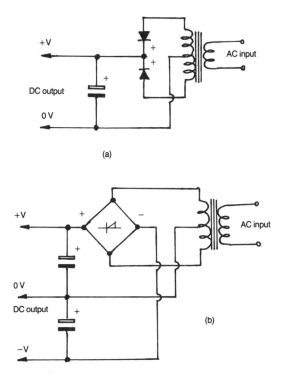

(a)

(b)

Fig. 5.18 *Simple DC power supplies.*

kind shown in Figs 5.18(a) and (b), to provide the DC supply to the amplifier. This has an output voltage which will decrease under increasing load, and it is necessary to know the characteristics of this supply in order to be able to specify the no-load output voltage which will be necessary so that the output voltage on maximum load will still be adequate to provide the rated output power.

Again, it is probable that the power supply for any medium to high power amplifier will provide more than enough current to burn out an expensive LS system in the event of an output transistor failure in the power amplifier, so some output fuse will be required. Unfortunately, most fuseholders tend to have a variable contact resistance, which introduces an unwelcome uncertainty into the critical LS output circuit.

The presence of the mains frequency-related AC ripple on the supply lines from such a simple power supply circuit, which will become worse as the output power to the load is increased, means that the voltage amplification circuitry must be chosen to have an adequate rejection of the supply line ripple.

Also, due to the relatively high circuit impedance, signal components present on the supply lines will worsen the distortion characteristics of the circuit if they break through into the signal path, since in a Class B or AB system these components will be highly distorted.

Such signal intrusion can also cause unwanted inter-channel break-through, and encourage the use of entirely separate power supply systems. Such simple power supplies may carry hidden cost penalties, too, since the working voltage rating of the power transistors must be adequate to withstand the no-load supply potential, and this may demand the use of more expensive, high-voltage transistors having relatively less good current gain or high frequency characteristics.

Stabilised power supplies

If, on the other hand, an electronically stabilised power supply is used, the output voltage can be controlled so that it is the same at full output power as under quiescent current conditions, and largely independent of mains voltage. Moreover, the supply can have a 're-entrant' overload characteristic, so that, under either supply line or LS output short-circuit conditions, supply voltage can collapse to a safely low level, and a monitor circuit can switch off the supplies to the amplifier if an abnormal DC potential, indicating output transistor failure, is detected at the LS terminals. Such protection should, of course, be fail-safe.

Other advantages which accrue from the use of stabilised power supplies for audio amplifiers are the low inherent power supply line 'hum' level, making the need for a high degree of supply line ripple rejection a less important characteristic in the design of the small signal stages, and the very good channel-to-channel isolation, because of the low inherent output impedance of such systems.

The principal sonic advantages which this characteristic brings are a more 'solid' bass response, and a more clearly defined stereo image, because the supply line source impedance can be very low, typically of the order of 0.1 ohms. In a conventional rectifier/capacitor type of power supply, a reservoir capacitor of 80 000 μF would be required to achieve this value at 20 Hz. The inherent inductance of an electrolytic capacitor of this size would give it a relatively high impedance at higher frequencies, whereas the output impedance of the stabilised supply would be substantially constant.

The design of stabilised power supplies is beyond the scope of this chapter, but a typical unit, giving separate outputs for the power output and preceding voltage amplifier stages of a power amplifier, and offering both re-entrant and LS voltage offset protection, is shown in Fig. 5.19.

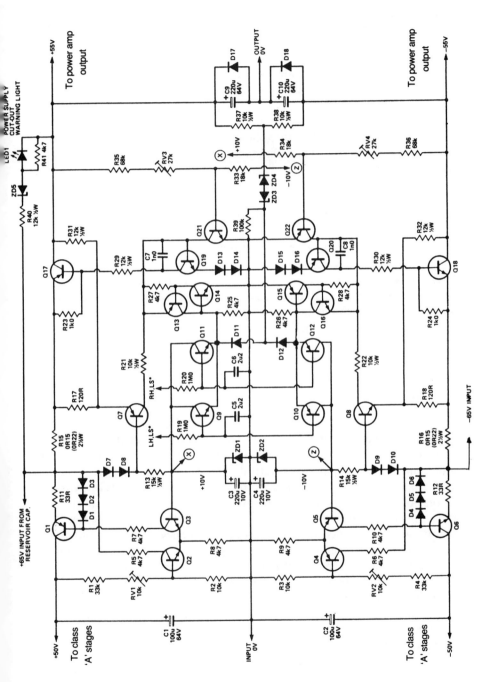

Fig. 5.19 *Twin dual-output stabilised power supply, with re-entrant overload protection and LS offset shut-down, by Linsley Hood.*

LS protection circuitry

This function is most commonly filled by the use of a relay, whose contacts in the output line from the amplifier to the LS unit are normally open, and only closed by the associated relay operating circuitry if all the appropriate conditions, including LS terminal DC offset and output load impedance, are within satisfactory limits.

Such circuitry is effective, but the relay contacts must have adequate contact area, and the relay must either be hermetically sealed or the contacts must be plated with gold or equivalent noble metal to preserve a low value of contact resistance in the presence of atmospheric contamination.

Output stage emitter follower configurations

Various circuit forms have been adopted for this purpose, of which the more useful ones are shown in Fig. 5.17. The actual form is known to have a bearing on the perceived sound quality of the design, and this was investigated by Otala and Lammasneimi (*Wireless World*, December 1980), who found that, with bipolar transistors, the symmetrical compound emitter follower circuit of Fig. 5.17(c) was significantly better than the complementary Darlington configuration of Fig. 5.17(a), in terms of the amplifier's ability to reject distortion components originating in load non-linearities from, or electromagnetic signal voltages originating in, the LS unit.

The actual desirable criterion in this case is the lowness of the output impedance of the output emitter follower configuration before overall negative feedback is applied to the amplifier, the point being that the feedback path within the amplifier is also a path whereby signals originating outside the amplifier can intrude within the signal path.

The major advantages offered by the use of power MOSFETs, typically in one or other of the configurations shown in Figs 5.17(e) or (f), are their greater intrinsic linearity in comparison with bipolar junction transistors, and their much higher transition frequency, which simplifies the design of a stable feedback amplifier having a low harmonic distortion. On the other hand, the output impedance of the simple source follower is rather high, and this demands a higher gain from the preceding voltage amplifier stage if an equivalent overall performance is to be obtained.

The inclusion of a low value resistor in the output circuit of the amplifier, between the emitter follower output and the LS load, as shown in the (1972) 75 W amplifier of Fig. 5.7, greatly reduces this type of load induced distortion and is particularly worth while in the case of the circuit layouts shown in Figs 5.17(a) and (e).

Power amplifier voltage gain stages

The general design systems employed in transistor gain stages have been examined in Chapter 4. However, for high quality audio power amplifiers higher open-loop stage gains, and lower inherent phase shift characteristics, will be required – to facilitate the use of large amounts of overall NFB to linearise output stage irregularities – than is necessary for the preceding small signal gain stages.

Indeed, with very many modern audio amplifier designs, the whole of the small signal pre-amplifier circuitry relies on the use of good quality integrated circuit operational amplifiers, of which there are a growing number which are pin compatible with the popular TL071 and TL072 single and dual FET-input op. amps. For power amplifier voltage stages, neither the output voltage nor the phase shift and large signal transient characteristics of such op. amps are adequate, so there has been much development of linear voltage gain circuitry, for the 'Class A' stages of power amplifiers, in which the principal design requirements have been good symmetry, a high gain/bandwidth product, a good transient response, and low-phase shift values within the audio range.

A wide range of circuit devices, such as constant current sources, current mirrors, active loads and 'long-tailed pairs' have been employed for this purpose, in many ingenious layouts. As a typical example, the circuit layout shown in Fig. 5.20, originally employed by National Semiconductors Inc. in its LH0001 operational amplifier, and adopted by Hitachi in a circuit recommended for use with its power MOSFETs, offers a high degree of symmetry, since Q_3/Q_4, acting as a current mirror, provide an active load equivalent to a symmetrically operating transistor amplifier, for the final amplifier transistor, Q_6.

This circuit offers a voltage gain of about 200 000 at low frequencies, with a stable phase characteristic and a high degree of symmetry. The derivation and development of this circuit was analysed by the author in *Wireless World* (July 1982).

An alternative circuit layout, of the type developed by Hafler, has been described by E. Borbely (*Wireless World*, March 1983), and is shown in Fig. 5.21. This is deliberately chosen to be fully symmetrical, so fas as the transistor characteristics will allow, to minimise any tendency to slew-rate limiting of the kind arising from stray capacitances charging or discharging through constant current sources. The open/loop gain is, however, rather lower than of the NS/Hitachi layout of Fig. 5.20.

Both unbypassed emitter resistors and base circuit impedance swamping resistors have been freely used in the Borbely design to linearise the transfer and improve the phase characteristics of the bipolar transistors used in this design, and a further improvement in the linearity of the output push–pull Darlington pairs ($Q_5/Q_6/Q_8/Q_9$) is obtained by the use of the 'cascode' connected buffer transistors Q_7 and Q_{10}.

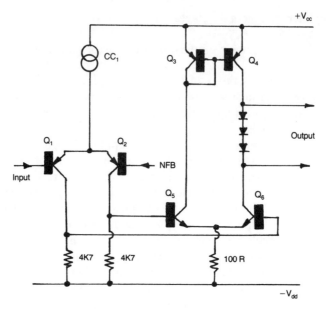

Fig. 5.20 *Symmetrical high gain stage.*

The particular merit of the cascode layout in audio circuitry is that the current flow through the cascode transistor is almost entirely controlled by the driver transistor in series with its emitter. In contrast, the collector potential of the driver transistor remains virtually constant, thus removing the deleterious effect of non-linear internal voltage dependent leakage resistances or collector-base capacitances from the driver device.

The very high degree of elaboration employed in recent high-quality Japanese amplifiers in the pursuit of improvements in amplifier performance, is shown in the circuit of the Technics SE−A100 voltage gain stage, illustrated in a somewhat simplified form in Fig. 5.22.

In this, an input long-tailed pair configuration, based on junction FETs (Q_1, Q_4 with CC_1), to take advantage of the high linearity of these devices, is cascode isolated (by Q_2, Q_3) from a current mirror circuit, (CM_1), which combines the output of the input devices in order to maximise the gain and symmetry of this stage, and drives a PNP Darlington pair amplifier stage (Q_5, Q_6).

The output transistor, Q_6, drives a current mirror (CM_2) through a cascode isolating transistor (Q_7) from Q_6 collector, and a further cascode isolated amplifier stage (Q_8, Q_9) from its emitter, for which the current mirror CM_2 serves as an active load. The amplified diode transistor, Q_{10}, serves to generate a DC offset potential, stabilised by a thermistor, (TH_1), to forward bias a succeeding push–pull pair of emitter followers.

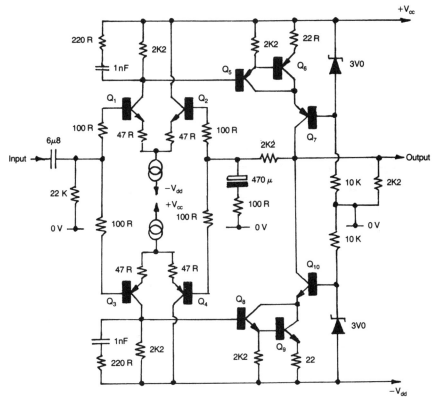

Fig. 5.21 *Symmetrical push–pull stage by Borbely.*

As a measure of the effectiveness of this circuit elaboration, the quoted harmonic distortion figures, for the whole amplifier, are typically of the order of 0.0002%.

ALTERNATIVE DESIGN APPROACHES

The fundamental problem in any 'Class B' or 'Class AB' transistor amplifier is that some non-linearity inevitably exists at the region where the current flow through one output transistor turns on and the other turns off.

This non-linearity can be minimised by the careful choice of output stage quiescent current, but the optimum performance of the amplifier depends on this current value being set correctly in the first place, and on its remaining constant at the set value throughout the working life of the amplifier.

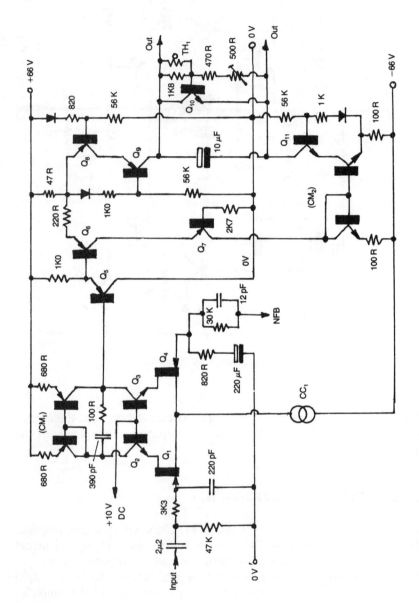

Fig. 5.22 *Technics voltage gain stage.*

One answer is, of course, to abandon 'Class AB' operation, and return to 'Class A', where both output transistors conduct during the whole AC output cycle, and where the only penalty for an inadvertent decrease in the operating current is a decrease in the maximum output power. The author's original four-transistor, 10 W 'Class A' design (*Wireless World*, April 1969) enjoys the distinction of being the simplest transistor operated power amplifier which is capable of matching the sound quality of contemporary valve designs. The problem, of course, is its limited power output.

The Blomley non-switching output circuit

The possibility of achieving a higher power 'Class AB' or even 'Class B' amplifier circuit, in which some circuit device is used to remove the fundamental non-linearity of the output transistor crossover region, in such circuits, is one which has tantalised amplifier designers for the past two decades, and various approaches have been explored. One of these which attracted a lot of interest at the time was that due to P. Blomley (*Wireless World*, February/March 1971), and which is shown, in simplified form, in Fig. 5.23.

In this, Q_1, Q_2, and Q_3 form a simple three-stage voltage amplifier, stabilised at HF by the inclusion of capacitor C_1 between Q_2 collector and Q_1 emitter. The use of a constant current load (CC_1) ensures good linearity

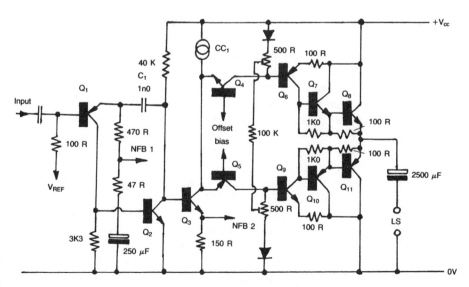

Fig. 5.23 *The Blomley non-switching push–pull output stage.*

from this stage. The crux of the design is the use of a pair of grounded base or cascode connected transistors, (Q_4, Q_5), whose bases are held, with a suitable DC offset between them, at some convenient mid-point DC level, which route the output current from the gain stage, Q_3, to one or other of the push–pull output triples (Q_6, Q_7, Q_8 and Q_9, Q_{10}, Q_{11}) which are arranged to have a significant current gain and also to be forward-biased, and therefore conducting, during the whole of the output signal cycle.

Although acclaimed as a non-switching 'Class AB' output configuration, in reality, the switching of the output half cycles of a 'Class B' system still takes place, but through the small signal transistors Q_4 and Q_5 which, since they are freed from the vagaries of the output loudspeaker load, and the need to pass a substantial output current, may be assumed to do the switching more cleanly and rapidly. Nevertheless, the need to maintain an accurate DC offset between the bases of these switching transistors still remains, and errors in this will worsen residual crossover distortion defects.

The Quad current dumping amplifier design

This unique and innovative circuit, subsequently employed commercially in the Quad 405 power amplifier, was first disclosed by P. J. Walker and M. P. Albinson at the fiftieth convention of the Audio Engineering Society, in the summer of 1975, and a technical description was given by Walker later in the year (*Wireless World*, December 1975).

This design claims to eliminate distortion due to the discontinuous switching characteristics of the unbiased, 'Class B', push–pull output transistor pair, by the use of a novel dual path feedback system, and thereby eliminate the need for precise setting-up of the amplifier circuit. It has been the subject of a very considerable subsequent analysis and debate, mainly hinging upon the actual method by which it works, and the question as to whether it does, or even whether it can, offer superior results to the same components used in a more conventional design.

What is not in doubt is that the circuit does indeed work, and that the requirement for correct adjustment of the output transistor quiescent current is indeed eliminated.

Of the subsequent discussion, (P. J. Baxandall, *Wireless World*, July 1976; Divan and Ghate, *Wireless World*, April 1977; Vanderkooy and Lipshitz, *Wireless World*, June 1978; M. McLoughlin, *Wireless World*, September/October 1983), the explanation offered by Baxandall is the most intellectually appealing, and is summarised below.

Consider a simple amplifier arrangement of the kind shown in Fig. 5.24(a), comprising a high-gain linear amplifier (A_1) driving an unbiased ('Class B') pair of power transistors (Q_1, Q_2), and feeding a load, Z_L.

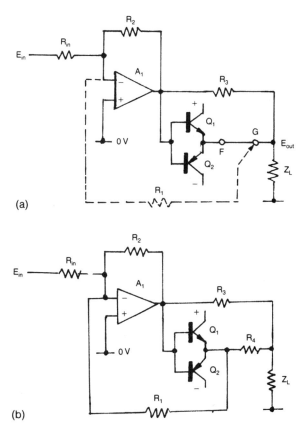

Fig. 5.24 *The basic current dumping system.*

Without any overall feedback the input/output transfer curve of this circuit would have the shape shown in the curve 'x' of Fig. 5.25, in which the characteristic would be steep from M' to N' while Q_2 was conducting, much flatter between N' and N while only the amplifier A_1 was contributing to the output current through the load, by way of the series resistance R_3, and then steeper again from N to M, while Q_1 was conducting.

If overall negative feedback is applied to the system via R_1, the extent of the discontinuity in the transfer curve can be made less, especially if the closed loop gain of A_1 is sufficiently high, leading to a more linear characteristic of the type shown in 'y'. However, it would still be unsatisfactory.

What is needed is some way of increasing the negative feedback applied to the system during the period in which Q_1 and Q_2 are conducting,

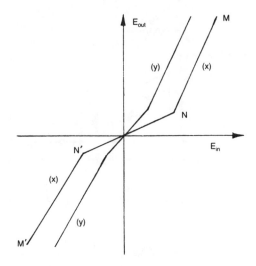

Fig. 5.25 *Current dumping amplifier transfer characteristics.*

to reduce the overall gain of the system so that the slope of the transfer characteristic of Fig. 5.25 is identical in the regions M' to N' and N to M to that between N' and N.

This can be done by inserting a small resistor, (R$_4$), between points F and G, in the output circuit of the push–pull emitter followers Q$_1$ and Q$_2$, so that there will be a voltage drop across this resistor, when Q$_1$ and Q$_2$ are feeding current into the load, and then deriving extra negative feedback from this point, (F), which will be related to the increased current flow into the load.

If the values of R$_1$, R$_2$, R$_3$ and R$_4$ are correctly chosen, in relation to the open loop gain of A$_1$, the distortion due to the unbiased output transistors will very nearly vanish, any residual errors being due solely to the imperfect switching characteristics of Q$_1$ and Q$_2$ and the phase errors at higher frequencies of the amplifier A$_1$.

Unfortunately, the output resistor, R$_4$, in the LS circuit would be wasteful of power, so Walker and Albinson substitute a small inductor for this component in the actual 404 circuit, and compensate for the frequency dependent impedance characteristics of this by replacing R$_2$ with a small capacitor.

While the amplifier circuit still works within the performance limits imposed by inevitable tolerance errors in the values of the components, this L and C substitution serves to complicate the theoretical analysis very considerably, and has led to a lot of the subsequent debate and controversy.

Feed-forward systems

The correction of amplifier distortion by the use of a comparator circuit, which would disclose the error existing between the input and output signals, so that the error could be separately amplified and added to or subtracted from the voltage developed across the load, was envisaged by H. S. Black, the inventor of negative feedback, in the earlier of his two US Patents (1686792/1928). Unfortunately, the idea was impracticable at that time because sufficiently stable voltage amplifiers were not obtainable.

However, if such a system could be made to work, it would allow more complete removal of residual errors and distortions than any more conventional negative feedback system, since with NFB there must always be some residual distortion at the output which can be detected by the input comparator and amplified to reduce the defect. In principle, the 'feed-forward' addition of a precisely measured error signal could be made to completely cancel the error, provided that the addition was made at some point, such as the remote end of the load, where it would not be sensed by the error detection system.

Two practical embodiments of this type of system by A. M. Sandman (*Wireless World*, October 1974), are shown in Figs 5.26(a) and (b). In the second, the iterative addition of the residual distortion components would allow the distortion in the output to be reduced to any level desired, while still allowing the use of a load circuit in which one end was connected to the common earth return line.

'Class S' amplifier systems

This ingenious approach, again due to A. M. Sandman (*Wireless World*, September 1982), has elements in common with the current dumping system, though its philosophy and implementation are quite different. The method employed is shown schematically in Fig. 5.27. In this a low-power, linear, 'Class A' amplifier, (A_1) is used to drive the load (Z_1) via the series resistor, R_4. The second linear amplifier, driving the 'Class B' push–pull output transistor pair, Q_1/Q_2, monitors the voltage developed across R_4, by way of the resistive attenuator, R_5/R_6, and adjusts the current fed into the load from Q_1 or Q_2 to ensure that the load current demand imposed on A_1 remains at a low level.

During the period in which neither Q_1 nor Q_2 is conducting, which will only be at regions close to the zero output level, the amplifier A_1 will supply the load directly through R_4. Ideally, the feedback resistor, R_2, should be taken to the top end of Z_1, rather than to the output of A_1, as shown by Sandman.

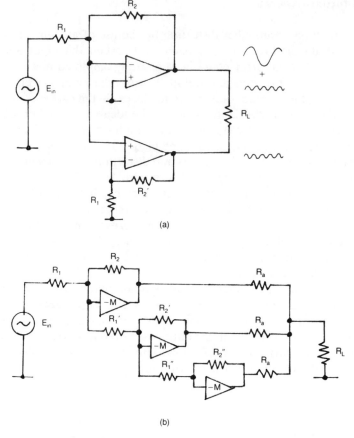

Fig. 5.26 *Feed-forward systems for reducing distortion.*

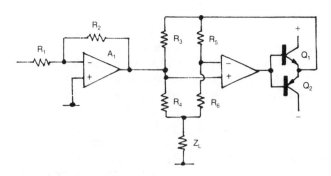

Fig. 5.27 *Sandman's 'Class S' system.*

A modified version of this circuit, shown in Fig. 5.28, is used by Technics in all its current range of power amplifiers, including the one for which the gain stage was shown in Fig. 5.22. In this circuit, a high-gain differential amplifier, (A_2), driving the current amplifier output transistors (Q_1, Q_2), is fed from the difference voltage existing between the outputs of A_1 and A_2, and the bridge balance control, R_y, is adjusted to make this value as close to zero as practicable.

Under this condition, the amplifier A_1 operates into a nearly infinite load impedance, as specified by Sandman, a condition in which its performance will be very good. However, because all the circuit resistances associated with R_3, R_4 and R_x are very low, if the current drive transistors are unable to accurately follow the input waveform, the amplifier A_1 will supply the small residual error current. This possible error in the operation of Q_1 and Q_2 is lessened, in the Technics circuit, by the use of a small amount of forward bias (and quiescent current) in transistors Q_1 and Q_2.

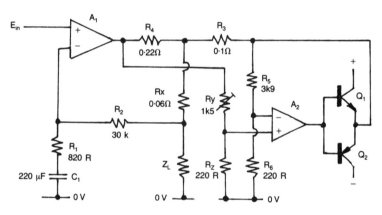

Fig. 5.28   *Technics power amplifier circuit (SE-A100).*

CONTEMPORARY AMPLIFIER DESIGN PRACTICE

This will vary according to the end use envisaged for the design. In the case of low cost 'music centre' types of system, the main emphasis will be upon reducing the system cost and overall component count. In such equipment, the bulk of the circuit functions, including the power output stages, will be handled by purpose built integrated circuits.

From the point of view of the manufacturers, these ICs and other specialised component groupings will be in-house items, only available from the manufacturer, and a more substantial profit margin on the sale of these to accredited repair agents will assist in augmenting the meagre profit levels imposed by competitive pressures on the original sale of the equipment.

In more prestigious equipment, intended to be assessed against similar units in the hi-fi market, the choice of circuit will lie between designs which are basically of the form shown in Fig. 5.14, but using more elaborate first stage amplifier circuitry, and with either bipolar or power MOSFET transistor output devices, or more elaborate systems derived from the Blomley, Sandman, or Current Dumping designs, or on systems in which the amplifier quiescent current is automatically adjusted during the output cycle with the aim of keeping the output stages operating, effectively, in 'Class A', but without the thermal dissipation penalty normally incurred by this.

Many, but not all, of the better quality units will employ stabilised DC power supplies, and virtually all of the high quality designs will be of the so-called direct-coupled form, in which the LS output is taken directly from the mid-point of the output emitter followers, without the interposition of a DC blocking output capacitor. (The use of true DC coupling from input to LS output is seldom found because of the problems of avoiding DC offset drift.)

Such direct-coupled amplifiers will, inevitably, employ symmetrical positive and negative supply lines, and in more up-market systems, the power supplies to the output stages will be separated from those for the preceding low power driver stages, and from any power supply to the preceding pre-amp. circuitry. This assists in keeping cross-channel break-through down to a low level, which is helpful in preserving the stability of the stereo image.

Great care will also be exercised in the best of contemporary designs in the choice of components, particularly capacitors, since the type of construction employed in these components can have a significant effect on sound quality. For similar reasons, circuitry may be chosen to mini-mise the need for capacitors, in any case.

Capacitors

Although there is a great deal of unscientific and ill-founded folklore about the influence of a wide variety of circuit components, from connecting wire to the nature of the fastening screws, on the final sound quality of an audio amplifying system, in the case of capacitors there is some technical basis for believing that imperfections in the operational characteristics of these components may be important, especially if such capacitors are used as an integral part of a negative feedback comparator loop.

The associated characteristics which are of importance include the inherent inductance of wound foil components, whether electrolytic or non-polar types, the piezo-electric or other electromechanical effects in the dielectric layer, particularly in ceramic components, the stored charge effects in some polymeric materials, of the kind associated with 'electret' formation, (the electrostatic equivalent of a permanent magnet, in which the material retains a permanent electrostatic charge), and non-linearities in the leakage currents or the capacitance as a function of applied voltage.

Polypropylene film capacitors, which are particularly valued by the subjective sound fraternity, because of their very low dielectric loss characteristics, are particularly prone to electret formation, leading to an asymmetry of capacitance as a function of polarising voltage. This effect is small, but significant in relation to the orders of harmonic distortion to which contemporary designs aspire.

Care in the decoupling of supply lines to the common earth return line is also of importance in the attainment of high performance, as is care in the siting and choice of earth line current paths. Such aspects of design care are not disclosed in the electronics circuit drawings.

SOUND QUALITY AND SPECIFICATIONS

Most of the performance specifications which relate to audio systems — such as the power output, (preferably measured as a continuous power into a specified resistive load), the frequency response, the input impedance and sensitivity, or the background noise level in comparison to some specified signal level — are reasonably intelligible to the non-technical user, and capable of verification on test.

However, the consideration which remains of major interest to the would-be user of this equipment is what it will sound like, and this is an area where it is difficult to provide adequate information from test measurements.

For example, it has been confidently asserted by well-known engineers that all competently designed power amplifiers operated within their ratings will sound alike. This may be true in respect of units from the same design stable, where the same balance of compromises has been adopted by the designer, but it is certainly untrue in respect of units having different design philosophies, and different origins.

As a particular case in point, until the mid-1970s a large majority of commercial designs employed a second-stage slew-rate limiting capacitor, in the mode discussed above, as a means of attaining stable operation without sacrifice of THD characteristics at the upper end of the audio band.

The type of sonic defect produced by slew-rate limiting is unattractive, and clearly audible by any skilled listener who has trained his ears to recognise the characteristic degradation of sound quality due to this. Since the publicity given to transient intermodulation distortion by Otala, this type of stabilisation is now seldom used in feedback amplifiers and other, technically more correct, methods are now more generally employed.

Since this type of shortcoming is now recognised, are we to accept that those of the preceding generation of designs which suffered from this defect (and which, in many cases, originated from the drawing boards of those same engineers who denied the existence of any differences) were, indeed, incompetently designed?

Design compromises

Unfortunately, the list of desirable parameters relating to the sound quality of audio amplifiers is a long one, and some of the necessary specifications are imperfectly understood. What is beyond doubt is that most of the designers operating in this field are well aware of their inability to attain perfection in all respects simultaneously, so that they must seek a compromise which will necessarily involve the partial sacrifice of perfection in one respect in order to obtain some improvement in some other mutually incompatible region.

The compromises which result, and which have an influence on the amplifier sound, are based on the personal judgement or preferences of the designer, and will vary from one designer to another.

An example of this is the case of low harmonic distortion figures at higher audio frequencies, and good transient performance and freedom from load induced instability, in a feedback amplifier. These characteristics are partially incompatible. However, 'THD' figures are prominently quoted and form an important part of the sales promotion literature and the reviewers report. Considerable commercial pressure therefore exists to attain a high performance in this respect.

Transient characteristics and feedback loop stability margins are not quoted, but shortcomings in either of these can give an amplifier a 'hard' or 'edgy' sound quality and it is not uncommon for poor amplifier transient performance to lead to a redistribution of energy in the time domain or the frequency spectrum which may amount to as much as a quarter of the total transient energy.

Bearing in mind the importance of good behaviour in this respect, it is to be regretted that if the transient performance of an amplifier is shown at all, it is likely to be shown only as a response to symmetrical square-waves, rather than to the more complex asymmetrical transients found in programme material.

Measurement systems

A measurement system which attempts to provide a more readily quantisable technique for assessing amplifier performance, in the hope of lessening the gap which exists between existing performance specifications – which mainly relate to steady state (i.e., sinusoidal) test signals – and the perceived differences in amplifier sound, has been devised by Y. Hirata (*Wireless World*, October 1981).

This technique uses asymmetrical, pulse type, input signals which approach more closely in form to the kinds of transient pressure waveforms generated by, for example, a bursting balloon, a hand clap, or cracker. The changes in these test waveforms caused by a variety of amplifier faults is shown by Hirata, but the interpretation of the results is too complex for it to be likely to replace the ubiquitous, if misleading, harmonic distortion figure as a criterion of amplifier goodness.

A further type of measurement, being explored by the BBC, has been described by R. A. Belcher (*Wireless World*, May 1978) using pseudo-random noise signals, derived by frequency shifting a 'comb filter' spectrum. This is claimed to give a good correlation with perceived sound quality, but is, again, too complex at the moment to offer an easily understood measurement by which a potential customer could assess the likely quality of an intended purchase.

Conclusions

The conclusion which can be drawn from this discussion is that harmonic distortion figures, on their own, offer little guidance about sound quality, except in a negative sense – that poor THD figures, in the 'worse than 0.5%' category, are likely to lead to poor sound quality. Fortunately, the understanding by design engineers of the requirements for good sound quality is increasing with the passage of time, and the overall quality of sound produced, even by 'budget' systems, is similarly improving.

In particular, there is now a growing appreciation of the relationship between the phase/frequency characteristics of the amplifier and the sound-stage developed by a stereo system of which it is a part.

There still seems to be a lot to be said for using the simplest and most direct approach, in engineering terms, which will achieve the desired end result – in that components which are not included will not fail, nor will they introduce any subtle degradation of the signal because of the presence of minor imperfections in their mode of operation. Also, simple systems are likely to have a less complex phase/frequency pattern than more highly elaborated circuitry.

For the non-technical user, the best guarantee of satisfaction is still a combination of trustworthy recommendation with personal experience, and the slow progress towards some simple and valid group of performance specifications, which would have a direct and unambiguous relationship to perceived sound quality, is not helped either by the advertisers frequent claims of perfection or the prejudices, sycophancy and favouritism of some reviewers.

On the credit side, the presence of such capricious critics, however unwelcome their views may be to those manufacturers not favoured with their approval, does provide a continuing stimulus to further development, and a useful counter to the easy assumption that because some aspect of the specification is beyond reproach the overall sound quality will similarly be flawless.

CHAPTER 6

The compact disc and digital audio

The only contact most users of audio equipment will have with digital signal processing (DSP) techniques is likely to be when they play a compact disc – the ubiquitous 'CD' – though there are a number of other 'digital' procedures in widespread use, such as in the transmission and decoding of FM stereo broadcasts, quite apart from the manufacture of CDs and other recordings, and it is probable that even more will be introduced in the next few years.

However, this is a very large field, which could not be covered adequately in a single chapter, so I propose to consider in detail only those techniques used in the initial encoding and subsequent replay of CDs, on the grounds that the basic electronic techniques used for this purpose have enough in common with those employed (for example in digital tape recording with both fixed head and rotary head machines) that an explanation of how the CD system works should also help to throw light on other similar processes.

WHY USE DIGITAL TECHNIQUES?

In the past fifty years, the equipment and technology needed for the production of high quality 'analogue' tape recordings (by which I mean recordings of signals which exist in 'real time' and are continuously variable) by way of multi-track, reel-to-reel machines, usually operating at a tape speed of 38.1 cms/s (15 in./s) has reached a very high standard of quality, and one might reasonably wonder why it should be thought necessary to seek to improve on this.

However, in practice, although the first-generation 'master tape' produced from a recording session using such a machine may be of superb technical quality, this will inevitably contain recording or performance faults which will require editing, and this edited master tape must then be copied – most probably twice or more times – in order to get a tidy version of the original. For security of the original recording, further working copies must then be made before the production of say, a number of vinyl discs, or compact cassettes, can begin.

Unfortunately, each time a successive analogue tape copy is made, some degradation of the original signal will occur, in respect of bandwidth and signal-to-noise ratio, and as a result of minor tape malfunctions and 'dropouts'. So by the time the recording has been converted into a cassette tape, or an undulating groove on the surface of a vinyl disc, a lot of the immediacy and transparency of the original recording will have been lost.

By comparison with this, a 'digital' recording – in which the analogue signal has been converted into a 'digitally encoded' electronic equivalent where the continuously variable voltage levels of the original signal are represented by a repetitively sampled sequence of alternating '0's and '1's, which signify clearly defined, constant and distinct electrical voltage levels – is, at least in principle, capable of being copied over and over again, without any degradation at all. Any minor errors in the received '0' or '1' levels can be automatically corrected, and freed from any spurious noise – a process which is obviously impracticable with any signal in analogue form.

In addition, the incoming signal, once converted into its digital form, need no longer exist in any specific time domain. After all, it is now just a collection of data, divided into a sequence of blocks. This allows the data to be divided, stored and manipulated, and reassembled in any way necessary for the purposes of recording or handling. It also means that, once the signal is converted into digital form, it is intrinsically free from any added 'rumble', 'wow' or 'flutter' or other intrusions due to the speed irregularities of the subsequent recording or replay systems. However, there are also snags.

PROBLEMS WITH DIGITAL ENCODING

Quantisation noise

Although a number of ways exist by which an analogue signal can be converted into its digital equivalent, the most popular, and the technique used in the CD, is the one known as 'pulse code modulation', usually referred to as 'PCM'. In this, the incoming signal is sampled at a sufficiently high repetition rate to permit the desired audio bandwidth to be achieved. In practice this demands a sampling frequency somewhat greater than twice the required maximum audio frequency. The measured signal voltage level, at the instant of sampling, is then represented numerically as its nearest equivalent value in binary coded form (a process which is known as 'quantisation').

This has the effect of converting the original analogue signal, after encoding and subsequent decoding, into a voltage 'staircase' of the kind shown in Fig. 6.1. Obviously, the larger the number of voltage steps in

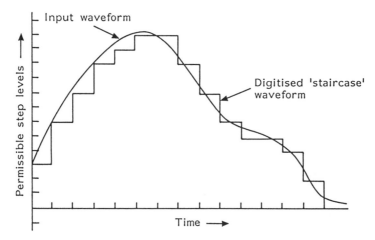

Fig. 6.1 *Digitally encoded/decoded waveform.*

which the analogue signal can be stored in digital form (that shown in the figure is encoded at '4-bit' – 2^4 or 16 possible voltage levels), the smaller each of these steps will be, and the more closely the digitally encoded waveform will approach the smooth curve of the incoming signal.

The difference between the staircase shape of the digital version and the original analogue waveform causes a defect of the kind shown in Fig. 6.2, known as 'quantisation error', and since this error voltage is not directly related in frequency or amplitude to the input signal it has many of the characteristics of noise, and is therefore also known as 'quantisation noise'. This error increases in size as the number of encoding levels is reduced. It will be audible if large enough, and is the first problem with digitally encoded signals. I will consider this defect, and the ways by which it can be minimised, later in this chapter.

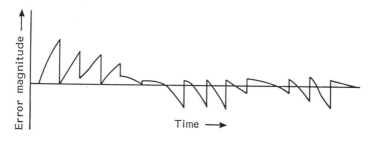

Fig. 6.2 *Quantisation error.*

Bandwidth

The second practical problem is that of the bandwidth which is necessary to store or transmit such a digitally encoded signal. In the case of the CD, the specified audio bandwidth is 20 Hz to 20 kHz, which requires a sampling frequency somewhat greater than 40 kHz. In practice, a sampling frequency of 44.1 kHz is used. In order to reduce the extent of the staircase waveform quantisation error, a 16-bit sampling resolution is used in the recording of the CD, equivalent to 2^{16} or 65 536 possible voltage steps. If 16 bits are to be transmitted in each sampling interval, then, for a stereo signal, the required bandwidth will be $2 \times 16 \times 44\,100$ Hz, or 1.4112 mHz, which is already seventy times greater than the audio bandwidth of the incoming signal. However, in practice, additional digital 'bits' will be added to this signal for error correction and other purposes, which will extend the required bandwidth even further.

Translation non-linearity

The conversion of an analogue signal both into and from its binary coded digital equivalent carries with it the problem of ensuring that the magnitudes of the binary voltage steps are defined with adequate precision. If, for example, '16-bit' encoding is used, the size of the 'most significant bit' (MSB) will be 32 768 times the size of the 'least significant bit' (LSB). If it is required that the error in defining the LSB shall be not worse than $\pm 0.5\%$, then the accuracy demanded of the MSB must be at least within $\pm 0.0000152\%$ if the overall linearity of the system is not to be degraded.

The design of any switched resistor network, for encoding or decoding purposes, which demanded such a high degree of component precision would be prohibitively expensive, and would suffer from great problems as a result of component ageing or thermal drift. Fortunately, techniques are available which lessen the difficulty in achieving the required accuracy in the quantisation steps. The latest technique, known as 'low bit' or 'bit-stream' decoding, side-steps the problem entirely by effectively using a time-division method, since it is easier to achieve the required precision in time, rather than in voltage or current, intervals.

Detection and correction of transmission errors

The very high bandwidths which are needed to handle or record PCM encoded signals means that the recorded data representing the signal must be very densely packed. This leads to the problem that any small blemish on the surface of the CD, such as a speck of dust, or a scratch, or a thumb print could blot out, or corrupt, a significant part of the information needed to reconstruct the original signal. Because of this, the real-life practicability of all digital record/replay systems will depend on the effectiveness of electronic techniques for the detection, correction or, if the worst comes to the worst, masking of the resultant errors. Some very sophisticated systems have been devised, which will also be examined later.

Filtering for bandwidth limitation and signal recovery

When an analogue signal is sampled, and converted into its PCM encoded digital equivalent, a spectrum of additional signals is created, of the kind shown in Fig. 6.3(a), where f_s is the sampling frequency and f_m is the upper modulation frequency. Because of the way in which the sampling process operates, it is not possible to distinguish between a signal having a frequency which is somewhat lower than half the sampling frequency and one which is the same distance above it; a problem which is called 'aliasing'. In order to avoid this, it is essential to limit the bandwidth of the incoming signal to ensure that it contains no components above $f_s/2$.

If, as is the case with the CD, the sampling frequency is 44.1 kHz, and the required audio bandwidth is 20 Hz to 20 kHz, +0/−1 dB, an input 'anti-aliasing' filter must be employed to avoid this problem. This filter must allow a signal magnitude which is close to 100% at 20 kHz, but nearly zero (in practice, usually −60 dB) at frequencies above 22.05 kHz. It is possible to design a steep-cut, low-pass filter which approximates closely to this characteristic using standard linear circuit techniques, but the phase shift, and group delay (the extent to which signals falling within the affected band will be delayed in respect of lower frequency signals) introduced by this filter would be too large for good audio quality or stereo image presentation.

This difficulty is illustrated by the graph of Fig. 6.4, which shows the relative group delay and phase shift introduced by a conventional low-pass analogue filter circuit of the kind shown in Fig. 6.5. The circuit shown gives only a modest −90 dB/octave attenuation rate, while the actual slope necessary for the required anti-aliasing characteristics (say, 0 dB at 20 kHz and −60 dB at 22.05 kHz) would be −426 dB/octave. If a group of filters of the kind shown in Fig. 6.5 were connected in series to increase the attenuation rate from −90 dB to −426 dB/octave, this would cause a group delay, at 20 kHz, of about 1 ms with respect to 1 kHz, and a relative phase shift of some 3000°, which would be clearly audible.*

---

*In the recording equipment it is possible to employ steep-cut filter systems in which the phase and group delay characteristics are more carefully controlled than would be practicable in a mass-produced CD replay system where both size and cost must be considered.

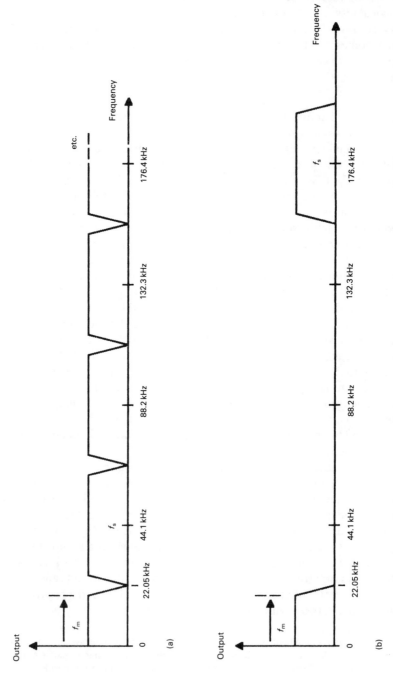

Fig. 6.3 *PCM frequency spectrum (a) when sampled at 44.1 kHz and (b) when four times oversampled.*

Similarly, since the frequency spectrum produced by a PCM encoded 20 kHz bandwidth audio signal will look like that shown in Fig. 6.3(a), it is necessary, on replay, to introduce yet another equally steep-cut low-pass filter to prevent the generation of spurious audio signals which would result from the heterodyning of signals equally disposed on either side of $f_s/2$.

An improved performance in respect both of relative phase error and of group delay in such 'brick wall' filters can be obtained using so-called 'digital' filters, particularly when combined with pre-filtering phase correction. However, this problem was only fully solved, and then only on replay (because of the limitations imposed by the original Philips CD patents), by the use of 'over-sampling' techniques, in which, for example, the sampling frequency is increased to 176.4 kHz ('four times over-sampling'), which moves the aliasing frequency from 22.05 kHz up to 154.35 kHz, giving the spectral distribution shown in Fig. 6.3(b). It is then a relatively easy matter to design a filter, such as that shown in Fig. 6.14, having good phase and group delay characteristics, which has a transmission near to 100% at all frequencies up to 20 kHz, but near zero at 154.35 kHz.

THE RECORD–REPLAY SYSTEM

The recording system layout

How the signal is handled, on its way from the microphone or other signal source to the final CD, is shown in the block diagram of Fig. 6.6. Assuming the signal has by now been reduced to a basic L–R stereo pair, this is amplitude limited to ensure that no signals greater than the possible encoding amplitude limit are passed on to the analogue-to-digital converter (ADC) stage. These input limiter stages are normally crosslinked in operation to avoid disturbance of the stereo image position if the maximum permitted signal level is exceeded, and the channel gain reduced in consequence of this, in only a single channel.

The signal is then passed to a very steep-cut 20 kHz anti-aliasing filter (often called a 'brick wall filter') to limit the bandwidth offered for encoding. This bandwidth limitation is a specific requirement of the digital encoding/decoding process, for the reasons already considered. It is necessary to carry out this filtering process after the amplitude limiting stage, because it is possible that the action of peak clipping may generate additional high frequency signal components. This would occur because 'squaring-off' the peaks of waveforms will generate a Fourier series of higher frequency harmonic components.

The audio signal, which is still, at this stage, in analogue form, is then passed to two parallel operating 16-bit ADCs, and, having now been converted into a digital data stream, is fed into a temporary data-storage

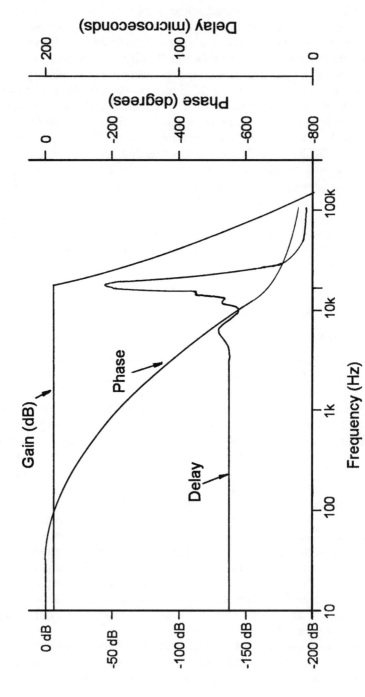

Fig. 6.4 *Responses of low-pass LC filter*

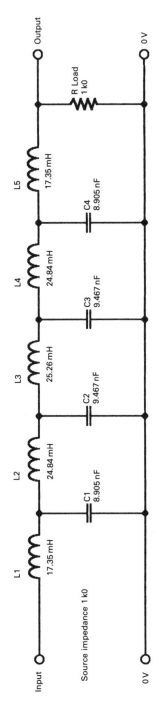

Fig. 6.5 *Steep-cut LP filter circuit.*

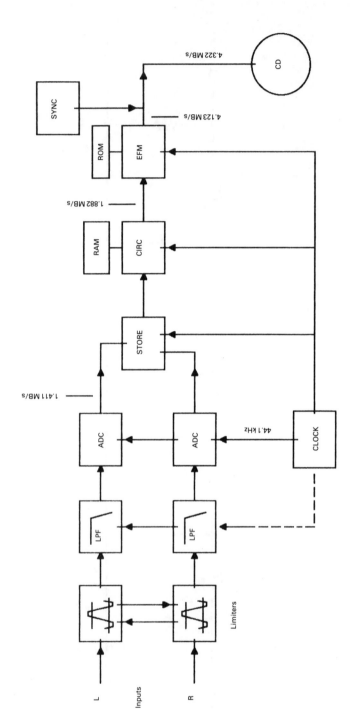

Fig. 6.6 *Basic CD recording system.*

device – usually a 'shift register' – from which the output data stream is drawn as a sequence of 8-bit blocks, with the 'L' and 'R' channel data now arranged in a consecutive but interlaced time sequence.

From the point in the chain at which the signal is converted into digitally encoded blocks of data, at a precisely controlled 'clock' frequency, to the final transformation of the encoded data back into analogue form, the signal is immune to frequency or pitch errors as a result of motor speed variations in the disc recording or replay process.

The next stage in the process is the addition of data for error correction purposes. Because of the very high packing density of the digital data on the disc, it is very likely that the recovered data will have been corrupted to some extent by impulse noise or blemishes, such as dust, scratches, or thumb prints on the surface of the disc, and it is necessary to include additional information in the data code to allow any erroneous data to be corrected. A number of techniques have been evolved for this purpose, but the one used in the CD is known as the 'Cross-interleave Reed–Solomon code' or CIRC. This is a very powerful error correction method, and allows complete correction of faulty data arising from quite large disc surface blemishes.

Because all possible '0' or '1' combinations may occur in the 8-bit encoded words, and some of these would offer bit sequences which were rich in consecutive '0's or '1's, which could embarrass the disc speed or spot and track location servo-mechanisms, or, by inconvenient juxtaposition, make it more difficult to read the pit sequence recorded on the disc surface, a bit-pattern transformation stage known as the 'eight to fourteen modulation' (EFM) converter is interposed between the output of the error correction (CIRC) block and the final recording. This expands the recorded bit sequence into the form shown in Fig. 6.7, to facilitate the operation of the recording and replay process. I shall explain the functions and method of operation of all these various stages in more detail later in this chapter.

Disc recording

This follows a process similar to that used in the manufacture of vinyl EP and LP records, except that the recording head is caused to generate a spiral pattern of pits in an optically flat glass plate, rather than a spiral groove in a metal one, and that the width of the spiral track is very much smaller (about 1/60th) than that of the vinyl groove. (Detail of the CD groove pattern is, for example, too fine to be resolved by a standard optical microscope.) When the master disc is made, 'mother' and 'daughter' discs are then made preparatory to the production of the stampers which are used to press out the track pattern on a thin (1.4 mm) plastics sheet, prior to the metallisation of the pit pattern for optical read-out in the final disc.

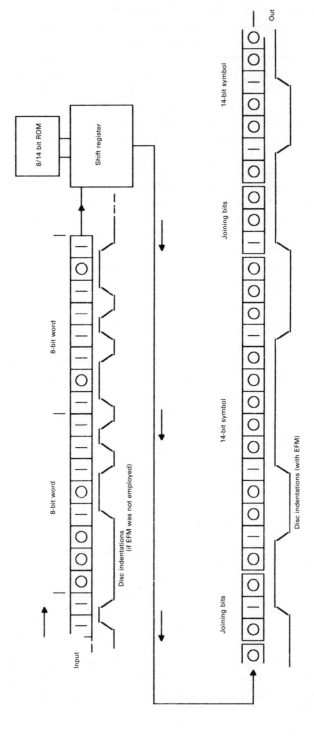

Fig. 6.7 *The EFM process.*

THE REPLAY SYSTEM

Physical characteristics

For the reasons shown above, the minimum bandwidth required to store the original 20 Hz to 20 kHz stereo signal in digitally encoded form has now been increased 215 fold, to some 4.3 MHz. It is, therefore, no longer feasible to use a record/replay system based on an undulating groove formed on the surface of a vinyl disc, because the excursions in the groove would be impractically close together, unless the rotational speed of the disc were to be enormously increased, which would lead to other problems such as audible replay noise, pick-up tracking difficulties, and rapid surface wear.

The technique adopted by Philips/Sony in the design of the CD replay system is therefore based on an optical pick-up mechanism, in which the binary coded '0's and '1's are read from a spiral sequence of bumps on an internal reflecting layer within a rapidly rotating (approx. 400 rpm) transparent plastic disc. This also offers the advantage that, since the replay system is non-contacting, there is no specific disc wear incurred in the replay of the records, and they have, in principle, if handled carefully, an indefinitely long service life.

CD PERFORMANCE AND DISC STATISTICS
Bandwidth   20 Hz–20 kHz, ±0.5 dB
Dynamic range   >90 dB
S/N ratio   >90 dB
Playing time (max.)   74 min
Sampling frequency   44.1 kHz
Binary encoding accuracy   16-bit (65 536 steps)
Disc diameter   120 mm
Disc thickness   1.2 mm
Centre hole diameter   15 mm
Permissible disc eccentricity (max.)   ±150 μm
Number of tracks (max.)   20 625
Track width   0.6 μm
Track spacing   1.6 μm
Tracking accuracy   ±0.1 μm
Accuracy of focus   ±0.5 μm
Lead-in diameter   46 mm
Lead-out diameter   116 mm
Track length (max.)   5300 m
Linear velocity   1.2–1.4 m/s

ADDITIONAL DATA ENCODED ON DISC
- *Error correction data.*
- *Control data*   Total and elapsed playing times, number of tracks, end of playing area, pre-emphasis,* etc.
- *Synchronisation signals*   added to define beginning and end of each data block
- *Merging bits*   used with EFM.

THE OPTICAL READ-OUT SYSTEM

This is shown, schematically, in Fig. 6.8, and consists of an infra-red laser light source, (GaAlAs, 0.5 mW, 780 nm), which is focused on a reflecting layer buried about 1 mm beneath the transparent 'active' surface of the disc being played. This metallic reflecting layer is deformed in the recording process, to produce a sequence of oblong humps along the spiral path of the recorded track (actually formed by making pits on the reverse side of the disc prior to metallisation). Because of the shallow depth of focus of the lens, due to its large effective numerical aperture ($f/0.5$) and the characteristics of the laser light focused upon the reflecting surface, these deformations of the surface greatly diminish the intensity of the incident light reflected to the receiver photocell, in comparison with that from the flat mirror-like surface of the undeformed disc. This causes the intensity of the light reaching the photocell to fluctuate as the disc rotates and causes the generation of the high speed sequence of electrical '0's and '1's required to reproduce the digitally encoded signal.

The signals representing '1's are generated by a photocell output level transition, either up or down, while '0's are generated electronically within the system by the presence of a timing impulse which is not coincident with a received '1' signal. This confers the valuable feature that the system defaults to a '0' if a data transition is not read, and such random errors can be corrected with ease in the replay system.

It is necessary to control the position of the lens, in relation both to the disc surface and to the recorded spiral sequence of surface lumps, to a high degree of accuracy. This is done by high speed closed-loop servo-mechanism systems, in which the vertical and lateral position of the whole optical read-out assembly is precisely adjusted by electro-mechanical actuators which are caused to operate in a manner which is very similar to the voice coil in a moving coil loudspeaker.

Two alternative arrangements are used for positioning the optical read-out assembly, of which the older layout employs a sled-type

---

*Note. Pre-emphasis may be added using either 15 μs (10 610 Hz) or 50 μs (3183 Hz) time constants.

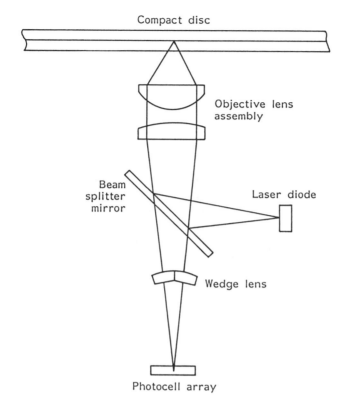

Fig. 6.8 *Single-beam optical read-out system.*

arrangement which moves the whole unit in a rectilinear manner across the active face of the disc. This maintains the correct angular position of the head, in relation to the recorded track, necessary when a 'three-beam' track position detector is used. Recent CD replay systems more commonly employ a single-beam lateral/vertical error detection system. Since this is insensitive to the angular relationship between the track and the head, it allows a simple pivoted arm structure to be substituted for the rectilinear-motion sled arrangement. This pivoted arm layout is cheaper to produce, is less sensitive to mechanical shocks, and allows more rapid scanning of the disc surface when searching for tracks.

Some degree of immunity from read-out errors due to scratches and dust on the active surface of the disc is provided by the optical characteristics of the lens which has a sufficiently large aperture and short focal length that the surface of the disc is out of focus when the lens is accurately focused on the plane of the buried mirror layer.

Electronic characteristics

The electronic replay system follows a path closely similar to that used in the encoding of the original recorded signal, though in reverse order, and is shown schematically in Fig. 6.9. The major differences between the record and replay paths are those such as 'oversampling', 'digital filtering' and 'noise shaping', intended to improve the accuracy of, and reduce the noise level inherent in, the digital-to-analogue transformation.

Referring to Fig. 6.9, the RF electrical output of the disc replay photocell, after amplification, is fed to a simple signal detection system, which mutes the signal chain in the absence of a received signal, to ensure inter-track silence. If a signal is present, it is then fed to the 'EFM' (fourteen to eight) decoder stage where the interface and 'joining' bits are removed, and the signal is passed as a group of 8-bit symbols to the 'Cross-interleave Reed–Solomon code' (CIRC) error correction circuit, which permits a very high level of signal restoration.

An accurate crystal-controlled clock regeneration circuit then causes the signal data blocks to be withdrawn in correct order from a sequential memory 'shift register' circuit, and reassembled into precisely timed and numerically accurate replicas of the original pairs of 16-bit (left and right channel) digitally encoded signals. The timing information from this stage is also used to control the speed of the disc drive motor and ensure that the signal data are recovered at the correct bit rate.

The remainder of the replay process consists of the stages in which the signal is converted back into analogue form, filtered to remove the unwanted high-frequency components, and reconstructed, as far as possible, as a quantisation noise-free copy of the original input waveform. As noted above, the filtering, and the accuracy of reconstruction of this waveform, are greatly helped by the process of 'oversampling', in which the original sampling rate is increased, on replay, from 44.1 kHz to some multiple of this frequency, such as 176.4 kHz or even higher. This process can be done by a circuit in which the numerical values assigned to the signal at these additional sampling points are obtained by interpolation between the original input digital levels, and as a matter of convenience, the same circuit arrangement will also provide a steep-cut filter having a near-zero transmission at half the sampling frequency.

THE 'EIGHT TO FOURTEEN MODULATION' TECHNIQUE (EFM)
This is a convenient shorthand term for what should really be described as '8-bit to 14-bit encoding/decoding', and is done for considerations of mechanical convenience in the record/replay process. As noted above, the '1's in the digital signal flow are generated by transitions from low to high, or from high to low, in the undulations on the reflecting surface of the disc.

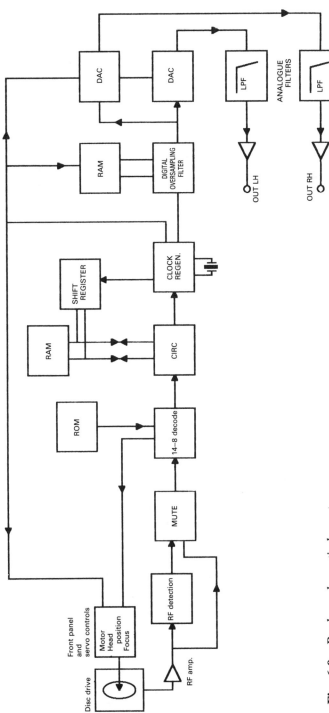

Fig. 6.9 *Replay schematic layout.*

On a statistical basis, it would clearly be possible, in an 8-bit encoded signal, for a string of eight or more '1's to occur in the bit sequence, the recording of which would require a rapid sequence of surface humps with narrow gaps between them, and this could be inconvenient to manufacture. Also, in the nature of things, these pits or humps will never have absolutely square, clean-cut edges, so the transitions from one sloping edge to another, where there is such a sequence of closely spaced humps, would also lead to a reduction in the replay signal amplitude, and might cause lost data bits.

On the other hand, a long sequence of '0's would leave the mirror surface of the disc unmarked by any signal modulation at all, and, bearing in mind the precise track and focus tolerances demanded by the replay system, this absence of signals at the receiver photocell would embarrass the control systems which seek to regulate the lateral and vertical position of the spot focused on the disc, and which use errors found in the bit repetition frequency, derived from the recovered sequence of '1's and '0's, to correct inaccuracies in the disc rotation speed. All these problems would be worsened in the presence of mechanical vibration.

The method chosen to solve this problem is to translate the 256-bit sequences possible with an 8-bit encoded signal into an alternative series of 256-bit sequences found in a 14-bit code, which are then reassembled into a sequence of symbols as shown graphically in Fig. 6.7. The requirements for the alternative code are that a minimum of two '0's shall separate each '1', and that no more than ten '0's shall occur in sequence. In the 14-bit code, it is found that there are 267 values which satisfy this criterion, of which 256 have been chosen, and stored in a ROM- based 'look-up' table. As a result of the EFM process, there are only nine different pit lengths which are cut into the disc surface during recording, varying from three to eleven clock periods in length.

Because the numerical magnitude of the output (EFM) digital sequence is no longer directly related to that of the incoming 8-bit word, the term 'symbol' is used to describe this or other similar groups of bits.

Since the EFM encoding process cannot by itself ensure that the junction between consecutive symbols does not violate the requirements noted above, an 'interface' or 'coupling' group of three bits is also added, at this stage, from the EFM ROM store, at the junction between each of these symbols. This coupling group will take the form of a '000', '100', '010', or '001' sequence, depending on the position of the '0's or '1's at the end of the EFM symbol. As shown in Fig. 6.6, this process increases the bit rate from 1.882 MB/s to 4.123 MB/s, and the further addition of uniquely styled 24-bit synchronising words to hold the system in coherence, and to mark the beginnings of each bit sequence, increases the final signal rate at the output of the recording chain to 4.322 MB/s. These additional joining and synchronising bits are stripped from the signal when the bit stream is decoded during the replay process.

DIGITAL-TO-ANALOGUE CONVERSION (DAC)

The transformation of the input analogue signal into, and back from, a digitally encoded bit sequence presents a number of problems. These stem from the limited time (22.7 μs) available for the conversion of each signal sample into its digitally encoded equivalent, and from the very high precision needed in allocating numerical values to each sample. For example, in a 16-bit encoded system the magnitude of the most significant bit (MSB) will be 32 768 times as large as the least significant bit (LSB). Therefore, to preserve the significance of a '0' to '1' transition in the LSB, both the initial and long-term precision of the electronic components used to define the size of the MSB would need to be better than ±0.00305%. (A similar need for accuracy obviously exists also in the ADC used in recording.)

Bearing in mind that even a 0.1% tolerance component is an expensive item, such an accuracy requirement would clearly present enormous manufacturing difficulties. In addition, any errors in the sizes of the steps between the LSB and the MSB would lead to waveform distortion during the encoding/decoding process: a distortion which would worsen as the signal became smaller.

Individual manufacturers have their own preferences in the choice of DAC designs, but a system due to Philips is illustrated, schematically, by way of example, in Fig. 6.10, an arrangement called 'dynamic element matching'. In this circuit, the outputs from a group of current sources, in a binary size sequence from 1 to 1/128, are summed by the amplifier A_1, whose output is taken to a simple 'sample and hold' arrangement to recover the analogue envelope shape from the impulse stream generated by the operation of the A_1 input switches (S_1–S_8). The required precision of the ratios between the input current sources is achieved by the use of switched resistor–capacitor current dividers each of which is only required to divide its input current into two equal streams.

Since in the CD replay process the input '16-bit' encoded signal is divided into two '8-bit' words, representing the MS and LS sections from e_1 to e_8 and from e_9 to e_{16}, these two 8-bit digital words can be separately D/A converted, with the outputs added in an appropriate ratio to give the final 16-bit D/A conversion.

DIGITAL FILTERING AND 'OVERSAMPLING'

It was noted above that Philips' original choice of sampling frequency (44.1 kHz) and of signal bandwidth (20 Hz to 20 kHz) for the CD imposed the need for steep-cut filtering both prior to the ADC and following the DAC stages. This can lead to problems caused by propagation delays and phase shifts in the filter circuitry which can degrade the sound quality. Various techniques are available which can lessen these problems, of which

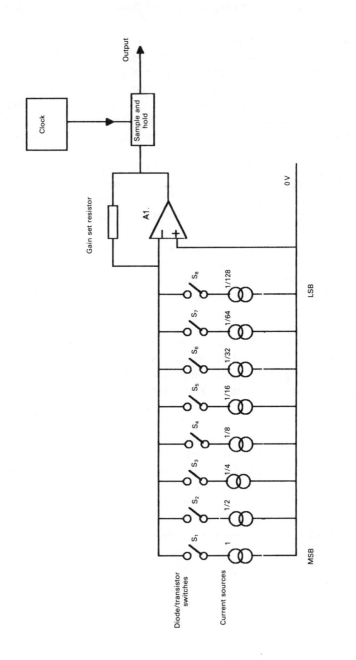

Fig. 6.10 *Dynamic matching DAC.*

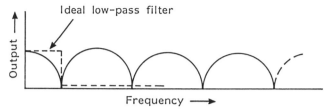

Fig. 6.11 *Comb filter frequency response.*

the most commonly used come under the headings of 'digital filtering' and 'oversampling'. Since these techniques are interrelated, I have lumped together the descriptions of both of these.

There are two practicable methods of filtering which can be used with digitally encoded signals. For these signals, use can be made of the effect that if a signal is delayed by a time interval, T_s, and this delayed signal is then combined with the original input, signal cancellation – partial or complete – will occur at those frequencies where T_s is equal to the duration of an odd number of half cycles of the signal. This gives what is known as a 'comb filter' response, shown in Fig. 6.11, and this characteristic can be progressively augmented to approach an ideal low-pass filter response (100% transmission up to some chosen frequency, followed by zero transmission above this frequency) by the use of a number of further signal delay and addition paths having other, carefully chosen, gain coefficients and delay times. (Although, in principle, this technique could also be used on a signal in analogue form, there would be problems in providing a non-distorting time delay mechanism for such a signal – a problem which does not arise in the digital domain.)

However, this comb filter type arrangement is not very conveniently suited to a system, such as the replay path for a CD, in which all operations are synchronised at a single specific 'clock' frequency or its sub-multiples, and an alternative digital filter layout, shown in Fig. 6.12 in simplified schematic form, is normally adopted instead. This provides a very steep-cut low-pass filter characteristic by operations carried out on the signal in its binary encoded digital form.

In this circuit, the delay blocks are 'shift registers', through which the signal passes in a 'first in, first out' sequence at a rate determined by the clock frequency. Filtering is achieved in this system by reconstructing the impulse response of the desired low-pass filter circuit, such as that shown in Fig. 6.13. The philosophical argument is that if a circuit can be made to have the same impulse response as the desired low-pass filter it will also have the same gain/frequency characteristics as that filter – a postulate which experiment shows to be true.

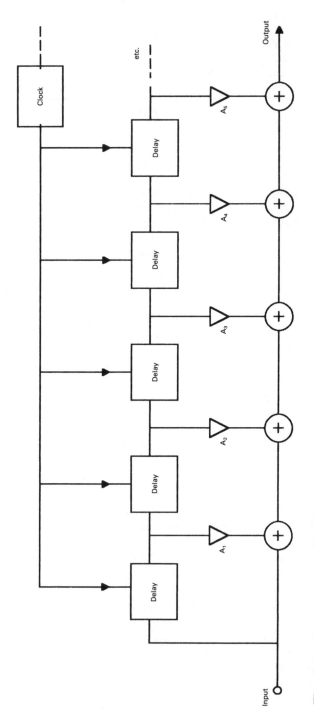

Fig. 6.12 *Basic oversampling filter.*

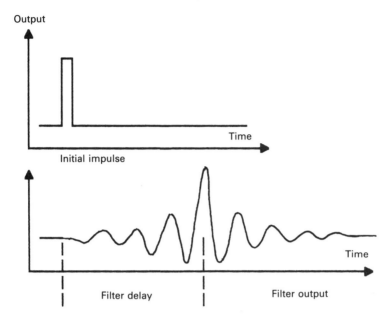

Fig. 6.13 *Impulse response of low-pass FIR filter. Zeros are $1/f_s$ apart; cut-off frequency $= f_s/2$.*

This required impulse response is built up by progressive additions to the signal as it passes along the input-to-output path, at each stage of which the successive delayed binary coded contributions are modified by a sequence of mathematical operations. These are carried out, according to appropriate algorithms, stored in 'look-up' tables, by the coefficient multipliers $A_1, A_2, A_3, \dots, A_n$. (The purpose of these mathematical manipulations is, in effect, to ensure that those components of the signal which recur more frequently than would be permitted by the notional 'cut-off' frequency of the filter will all have a coded equivalent to zero magnitude.) Each additional stage has the same attenuation rate as a single-pole RC filter (-6 dB/octave), but with a strictly linear phase characteristic, which leads to zero group delay.

This type of filter is known either as a 'transversal filter', from the way in which the signal passes through it, or a 'finite impulse response' (FIR) filter, because of the deliberate omission from its synthesised impulse response characteristics of later contributions from the coefficient multipliers. (There is no point in adding further terms to the $A_1, \dots, A_n$ series when the values of these operators tend to zero.)

Some contemporary filters of this kind use 128 sequential 'taps' to the transmission chain, giving the equivalent of a -768 dB/octave low-pass filter. This demonstrates, incidentally, the advantage of handling signals in

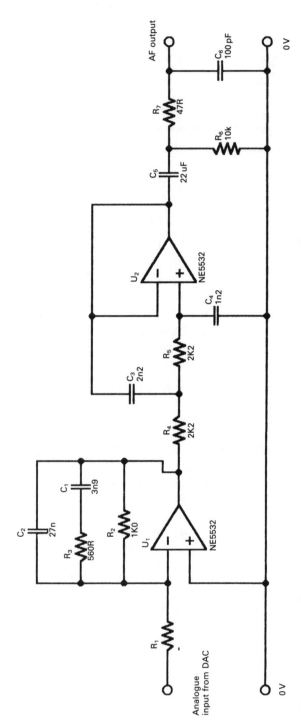

Fig. 6.14 *Linear phase LP filter.*

the digital domain, in that a 128-stage analogue filter would be very complex and also have an unacceptably high thermal noise background.

If the FIR clock frequency is increased to 176.4 kHz, the action of the shift registers will be to generate three further signal samples, and to interpolate these additional samples between those given by the original 44.1 kHz sampling intervals – a process which would be termed 'four times oversampling'.

The simple sample-and-hold stage, at the output of the DAC shown in Fig. 6.10, will also assist filtering since it will attenuate any signals occurring at the clock frequency, to an extent determined by the duration of the sampling operation – called the sampling 'window'. If the window length is near 100% of the cycle time the attenuation of the S/H circuit will be nearly total at f_s.

Oversampling, on its own, would have the advantage of pushing the aliasing frequency up to a higher value, which makes the design of the anti-aliasing and waveform reconstruction filter a much easier task to accomplish, using simple analogue-mode low-pass filters whose character-istics can be tailored so that they introduce very little unwanted group delay and phase shift. A typical example of this approach is the linear phase analogue filter design, shown in Fig. 6.14, used following the final 16-bit DACs in the replay chain.

However, the FIR filter shown in Fig. 6.12 has the additional effect of computing intermediate numerical values for the samples interpolated between the original 44.1 kHz input data, which makes the discontinuities in the PCM step waveform smaller, as shown in Fig. 6.15. This reduces the

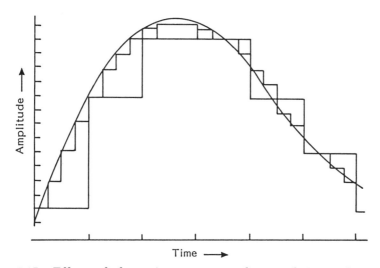

Fig. 6.15 *Effect of four times oversampling and interpolation of intermediate values.*

quantisation noise and also increases the effective resolution of the DAC. As a general rule, an increase in the replay sampling rate gives an improvement in resolution which is equivalent to that given by a similar increase in encoding level – such that a four times oversampled 14-bit decoder would have the same resolution as a straight 16-bit decoder.

Yet another advantage of oversampling is that it increases the bandwidth over which the 'quantisation noise' will be spread – from 22.05 kHz to 88.2 kHz in the case of a four times oversampling system. This reduces the proportion of the total noise which is now present within the audible (20 Hz to 20 kHz) part of the frequency spectrum – especially if 'noise shaping' is also employed. I will examine this aspect later in this chapter.

'DITHER'

If a high-frequency noise signal is added to the waveform at the input to the ADC, and if the peak-to-peak amplitude of this noise signal is equal to the quantisation step 'Q', both the resolution and the dynamic range of the converter will be increased. The reason for this can be seen if we consider what would happen if the actual analogue signal level were to lie somewhere between two quantisation levels. Suppose, for example, in the case of an ADC, that the input signal had a level of 12.4, and the nearest quantisation levels were 12 and 13. If dither had been added, and a sufficient number of samples were taken, one after another, there would be a statistical probability that 60% of these would be attributed to level 12, and 40% would be attributed to level 13, so that, on averaging, the final analogue output from the ADC/DAC process would have the correct value of 12.4.

A further benefit is obtained by the addition of dither at the output of the replay DACs (most simply contrived by allowing the requisite amount of noise in the following analogue low-pass filters) in that it will tend to mask the quantisation 'granularity' of the recovered signal at low bit levels. This defect is particularly noticeable when the signal frequency happens to have a harmonic relationship with the sampling frequency.

THE 'BITSTREAM' PROCESS AND 'NOISE SHAPING'

A problem in any analogue-to-digital or digital-to-analogue converter is that of obtaining an adequate degree of precision in the magnitudes of the digitally encoded steps. It has been seen that the accuracy required, in the most significant bit in a 16-bit converter, was better than 0.00305% if '0'–'1' transitions in the LSB were to be significant. Similar, though lower, orders of accuracy are required from all the intermediate step values. Achieving this order of accuracy in a mass-produced consumer article is difficult and expensive. In fact, differences in tonal quality between CD players are likely to be due, in part, to inadequate precision in the DACs.

As a means of avoiding the need for high precision in the DAC converters, Philips took advantage of the fact that an effective improvement in resolution could be achieved merely by increasing the sampling rate, and this could then be traded-off against the number of bits in the quantisation level. Furthermore, whatever binary encoding system is adopted, the first bit in the received 16-bit word must always be either a '0' or a '1', and in the 'two's complement' code used in the CD system, the transition in the MSB from '0' to '1' and back will occur at the mid-point of the input analogue signal waveform.

This means that if the remaining 15 bits of a 16-bit input word are stripped off and discarded, this action will have the effect that the input digital signal will have been converted – admittedly somewhat crudely – into a voltage waveform of analogue form. Now, if this '0/1' signal is 256 times oversampled, in the presence of dither, an effective 9-bit resolution will be obtained from two clearly defined and easily stabilised quantisation levels: a process for which Philips have coined the term 'Bitstream' decoding.

Unfortunately, such a low-resolution quantisation process will incur severe quantisation errors, manifest as a high background noise level. Philips' solution to this is to employ 'noise shaping', a procedure in which, as shown in Fig. 6.16, the noise components are largely shifted out of the 20 Hz to 20 kHz audible region into the inaudible upper reaches of the new 11.29 MHz bandwidth.

The proposition is, in effect, that a decoded digital signal consists of the pure signal, plus a noise component (caused by the quantisation error)

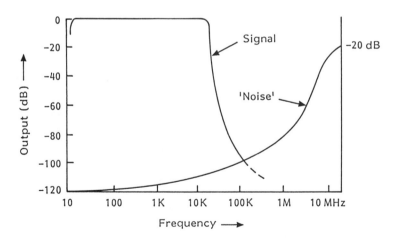

Fig. 6.16 *Signal/noise spectrum after 'noise-shaping'.*

related to the lack of resolution of the decoding process. It is further argued that if this noise component is removed by filtering what remains will be the pure signal – no matter how poor the actual resolution of the decoder. Although this seems an unlikely hypothesis, users of CD players employing the 'Bitstream' system seem to agree that the technique does indeed work in practice. It would seem therefore that the greater freedom from distortion, which could be caused by errors in the quantisation levels in high bit-level DACs, compensates for the crudity of a decoding system based on so few quantisation steps.

Mornington-West (*Newnes Audio and Hi-Fi Handbook* (2nd edition, p. 141) quotes oversampling values of 758 and 1024 times respectively for 'Technics' and 'Sony' 'low-bit' CD players, which would be equivalent in resolution to 10.5-bit and 11-bit quantisation if a simple '0' or '1' choice of encoding levels was used. Since the presence of dither adds an effective 1-bit to the resolution and dynamic range, the final figures would become 10-, 11.5- and 12-bit resolution respectively for the Philips, Technics and Sony CD players.

However, such decoders need not use the single-bit resolution adopted by Philips, and if a 2-bit or 4-bit quantisation was chosen as the base to which the oversampling process was applied – an option which would not incur significant problems with accuracy of quantisation – this would provide low-bit resolution values as good as the 16-bit equivalents, at a lower manufacturing cost and with greater reproducibility. Ultimately, the limit to the resolution possible with a multiple sampling decoder is set by the time 'jitter' in the switching cycles and the practicable operating speeds of the digital logic elements used in the shift registers and adders. In the case of the 1024 times oversampling 'Sony' system, a 44.1584 MHz clock speed is required, which is near the currently available limit.

ERROR CORRECTION

The possibility of detecting and correcting replay errors offered by digital audio techniques is possibly the largest single benefit offered by this process, because it allows the click-free, noise-free background level in which the CD differs so obviously from its vinyl predecessors. Indeed, were error correction not possible, the requirements for precision of the CD manufacturing and replay process would not be practicable.

Four possible options exist for the avoidance of audible signal errors, once these have been detected. These are: the replacement of the faulty word or group of words by correct ones: the substitution of the last correct word for the one found to be faulty – on the grounds that an audio signal is likely to change relatively slowly in amplitude in comparison with the 44.1 kHz sample rate: linear interpolation of intermediate sample values in the gaps caused by the deletion of incorrectly received words, and, if the worst comes to the worst, the muting of the signal for the duration of the error.

Of these options, the replacement of the faulty word, or group of words, by a correct equivalent is clearly the first preference, though it will, in practice, be supplemented by the other error-concealment techniques. The error correction system used in the CD replay process combines a number of error correction features and is called the 'Cross-interleave Reed–Solomon Code' (CIRC) system. It is capable of correcting an error of 3500 bits, and of concealing errors of up to 12 000 bits by linear interpolation. I will look at the CIRC system later, but, meanwhile, it will be helpful to consider some of the options which are available.

Error detection

The errors which are likely to occur in a digitally encoded replay process are described as 'random' when they affect single bits and 'burst' when they affect whole words or groups of words. Correcting random errors is easier, so the procedure used in the Reed–Solomon code endeavours to break down burst errors into groups of scattered random errors. However, it greatly facilitates remedial action if the presence and location of the error can be detected and 'flagged' by some added symbol.

Although the existence of an erroneous bit in an input word can sometimes be detected merely by noting a wrong word length, the basic method of detecting an error in received words is by the use of 'parity bits'. In its simplest form, this would be done by adding an additional bit to the word sequence, as shown in Fig. 6.17(a), so that the total (using the logic rules shown in the figure) always added up to zero (a method known as 'even parity'). If this addition had been made to all incoming words, the presence of a word plus parity bit which did not add up to '0' could be detected instantly by a simple computer algorithm, and it could then be rejected or modified.

Faulty bit/word replacement

Although the procedure shown above would alert the decoder to the fact that the word was in error, the method could not distinguish between an incorrect word and an incorrect parity bit – or even detect a word containing two separate errors, though this might be a rare event. However, the addition of extra parity bits can indeed correct such errors as well as detect them, and a way by which this could be done is shown in Fig. 6.17(b). If a group of four 4-bit input words, as shown in lines a–d, each

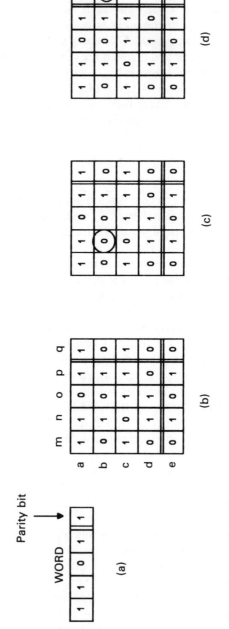

Fig. 6.17 *Parity bit error correction. Logic: $0 + 0 = 0$, $0 + 1 = 1$, $1 + 1 = 0$.*

has a parity bit attached to it, as shown in column q, so that each line has an even parity, and if each column has a parity bit attached to it, for the same purpose, as shown in line e, then an error, as shown in grid reference (b.n) in Fig. 6.17(c), could not only be detected and localised as occurring at the intersection of row b and column n, but it could also be corrected, since if the received value '0' is wrong, the correct alternative must be '1'.

Moreover, the fact that the parity bits of column q and row e both have even parity means that, in this example, the parity bits themselves are correct. If the error had occurred instead in one of the parity bits, as in Fig. 6.17(d), this would have been shown up by the fact that the loss of parity occurred only in a single row – not in both a row and a column.

So far, the addition of redundant parity bit information has offered the possibility of detecting and correcting single bit 'random' errors, but this would not be of assistance in correcting longer duration 'burst' errors, comprising one or more words. This can be done by 'interleaving', the name which is given to the deliberate and methodical scrambling of words, or the bits within words, by selectively delaying them, and then re-inserting them into the bit sequence at later points, as shown in Fig. 6.18. This has the effect of converting a burst error, after de-interleaving, into a scattered group of random errors, a type of fault which is much easier to correct.

A further step towards the correction of larger-duration errors can be made by the use of a technique known as 'cross-interleaving'. This is done by reassembling the scrambled data into 8-bit groups without descrambling. (It is customary to refer to these groups of bits as 'symbols' rather than words because they are unrelated to the signal.) Following this, these symbols are themselves mixed up in their order by removal and re-insertion at different delay intervals. In order to do this it is necessary to have large bit-capacity shift registers, as well as a fast microprocessor which can manipulate the information needed to direct the final descrambling sequences, and generate and insert the restored and corrected signal words.

To summarise, errors in signals in digital form can be corrected by a variety of procedures. In particular, errors in individual bits can be corrected by the appropriate addition of parity bits, and burst errors affecting words, or groups of words, can be corrected by interleaving and de-interleaving the signal before and after transmission – a process which separates and redistributes the errors as random bit faults, correctable by parity techniques.

A variety of strategies has been devised for this process, aimed at achieving the greatest degree of error removal for the lowest necessary number of added parity bits. The CIRC error correction process used for CDs is very efficient in this respect, since it only demands an increase in transmitted data of 33.3%, and yet can correct burst errors up to 3500 bits in length. It can conceal, by interpolation, transmission errors up to 12 000

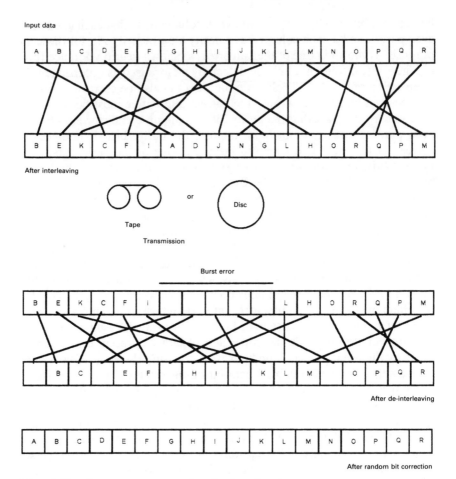

Fig. 6.18   *Burst error correction by interleaving.*

bits in duration – an ability which has contributed enormously to the sound quality of the CD player by comparison with the vinyl disc.

From the point of view of the CD manufacturer, it is convenient that the complete CIRC replay error correction and concealment package is available from several IC suppliers as part of a single large-scale integrated (LSI) chip. From the point of view of the serious CD user, it is preferable that the error correction system has to do no more work than it must, since although the errors will mainly be restored quite precisely, it may be necessary, sometimes, for the system to substitute approximate, interpolated values for the signal data, and the effect of frequent corrections may be audible to the critical listener. So treat CDs with care, keep them clean, and try to avoid surface scratches.

CHAPTER 7

Test instruments and measurements

Audio equipment is, by definition, ultimately intended to provide sounds which will be heard by a listener. In spite of rather more than a century of experience in the electrical transmission and reception of audible signals, the relationships between the electrical waveforms into which sound patterns are transformed and the sound actually heard by the listener are still not fully understood.

This situation is complicated by the observable fact that there is a considerable variation from person to person in sensitivity to, and preferences in respect of, sound characteristics – particularly where these relate to modifications or distortions of the sound. Also, since the need to describe or indeed the possibility of producing such modified or distorted sounds is a relatively new situation, we have not yet evolved a suitable and agreed vocabulary by which we can define our sensations.

There is, however, some agreement, in general, about the types of defect in electrical signals which, in the interests of good sound quality, the design engineer should seek to avoid or minimise. Of these, the major ones are those associated with waveform distortion, under either steady-state or transient conditions: or with the intrusion of unwanted signals: or with relative time delays in certain parts of the received signal in relation to others: or with changes in the pitch of the signal; known as 'wow' or 'flutter', depending on its frequency. However, the last kind of defect is only likely to occur in electro- mechanical equipment, such as turntables or tape drive mechanisms, used in the recording or replay of signals.

It is sometimes claimed that, at least so far as the design of purely electronic equipment is concerned, the performance can be calculated sufficiently precisely that it is unnecessary to make measurements on a completed design for any other reason than simply to confirm that the target specification is met. Similarly, it is argued that it is absurd to attempt to endorse or reject any standard of performance by carrying out listening trials. This is so since, even if the results of calculations were not adequate to define the performance, instrumental measurements are so much more sensitive and reproducible than any purely 'subjective' assessments that no significant error could escape instrumental detection.

Unfortunately, all these assertions remain a matter of some dispute. With regard to the first of these points – the need for instrumental measurements – the behaviour patterns of many of the components, both 'passive' and 'active', used in electronic circuit design are complex, particularly under transient conditions, and it may be difficult to calculate precisely what the final performance of any piece of audio equipment will be, over a comprehensive range of temperatures or of signal and load conditions. However, appropriate instrumental measurements can usually allow a rapid exploration of the system behaviour over the whole range of interest.

On the second point, the usefulness of subjective testing, the problem is to define just how important any particular measurable defect in the signal process is likely to prove in the ear of any given listener. So where there is any doubt, recourse must be had to carefully staged and statistically valid comparative listening trials to try to determine some degree of consensus. These trials are expensive to stage, difficult to set up and hard to purge of any inadvertent bias in the way they are carried out. They are therefore seldom done, and even when they are, the results are disputed by those whose beliefs are not upheld.

INSTRUMENT TYPES

An enormous range of instruments is available for use in the test laboratory, among which, in real-life conditions, the actual choice of equipment is mainly limited by considerations of cost, and of value for money in respect to the usefulness of the information which it can provide.

Although there is a wide choice of test equipment, much of the necessary data about the performance of audio gear can be obtained from a relatively restricted range of instruments, such as an accurately calibrated signal generator, with sinewave and square-wave outputs, a high input impedance, wide bandwidth AC voltmeter, and some instrument for measuring waveform distortion – all of which would be used in conjunction with a high-speed double trace cathode-ray oscilloscope. I have tried, in the following pages, to show how these instruments are used in audio testing, how the results are interpreted, and how they are made. Since some of the circuits which can be used are fairly simple, I have given details of the layouts needed so that they could be built if required by the interested user.

SIGNAL GENERATORS

Sinewave oscillators

Variable frequency sinusoidal input test waveforms are used for determining the voltage gain, the system bandwidth, the internal phase shift or group delay, the maximum output signal swing and the amount of waveform distortion introduced by the system under test. For audio purposes, a frequency range of 20 Hz to 20 kHz will normally be adequate, though practical instruments will usually cover a somewhat wider bandwidth than this. Except for harmonic distortion measurements, a high degree of waveform purity is probably unnecessary, and stability of output as a function of time and frequency is probably the most important characteristic for such equipment.

It is desirable to be able to measure the output signal swing and voltage gain of the equipment under specified load conditions. In, for example, an audio power amplifier, this would be done to determine the input drive requirements and output power which can be delivered by the amplifier. For precise measurements, a properly specified load system, a known frequency source and an accurately calibrated, RMS reading, AC voltmeter would be necessary, together with an oscilloscope to monitor the output waveform to ensure that the output waveform is not distorted by overloading.

Some knowledge of the phase errors (the relative time delay introduced at any one frequency in relation to another) can be essential for certain uses – for example, in long-distance cable transmission systems – but in normal audio usage such relative phase errors are not noticeable unless they are very large. This is because the ear is generally able to accept without difficulty the relative delays in the arrival times of sound pressure waves due to differing path lengths caused by reflections in the route from the speaker to the ear.

Oscillators designed for use with audio equipment will typically cover the frequency range 10 Hz to 100 kHz, with a maximum output voltage of, perhaps, 10 V RMS. For general purpose use, harmonic distortion levels in the range 0.5–0.05% will probably be adequate, though equipment intended for performance assessments on high quality audio amplifiers will usually demand waveform purity (harmonic distortion) levels at 1 kHz in the range from 0.02% down to 0.005%, or lower. In practice, with simpler instruments, the distortion levels will deteriorate somewhat at the high and low frequency ends of the output frequency band.

A variety of electronic circuit layouts have been proposed for use as sinewave signal generators, of which by far the most popular is the 'Wien Bridge' circuit shown in Fig. 7.1. It is a requirement for continuous oscillation in any system that the feedback from output to input shall have zero (or some multiple of 360°) phase shift at a frequency where the feedback loop gain is very slightly greater than unity, though to avoid waveform distortion, it is necessary that the gain should fall to unity at some value of output voltage within its linear voltage range.

In the Wien bridge, if $R_1 = R_2 = R$ and $C_1 = C_2 = C$, the condition for zero phase shift in the network is met when the output frequency, $f_0 = 1/(2\pi RC)$. At this frequency the attenuation of the RC network, from

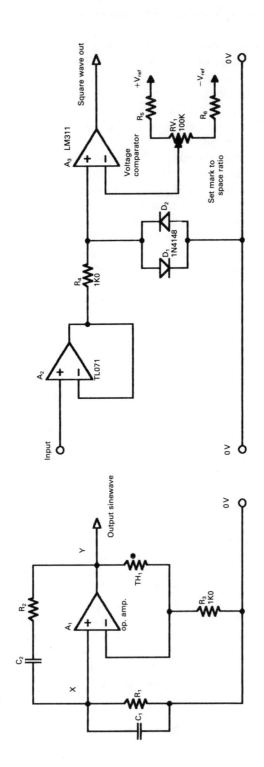

Fig. 7.1 *Basic Wien bridge oscillator and add-on square-wave generator.*

Y to X, in Fig. 7.1, is 1/3. The circuit shown will therefore oscillate at f_0 if the gain of the amplifier, A_1, is initially slightly greater than three times. The required gain level can be obtained automatically by the use of a thermistor (TH_1) in the negative feedback path, and the correct choice of the value of R_3.

Since C_1R_1 and C_2R_2 are the frequency-determining elements, the output frequency of the oscillator can be made variable by using a twin-gang variable resistor as R_1/R_2 or a twin gang capacitor as C_1/C_2. If a modern, very low distortion, operational amplifier, such as the LM833, the NE5534 or the OP27 is used as the amplifier gain block (A_1) in this circuit, the principle source of distortion will be that caused by the action of the thermistor (TH_1) used to stabilise the output signal voltage, where, at lower frequencies, the waveform peaks will tend to be flattened by its gain-reduction action. With an RS Components 'RA53' type thermistor as TH_1, the output voltage will be held at approximately 1 V RMS, and the THD at 1 kHz will be typically of the order of 0.008%.

The output of almost any sinewave oscillator can be converted into square-wave form by the addition of an amplifier which is driven into clipping. This could be either an op. amp., or a string of CMOS inverters, or, preferably, a fast voltage comparator IC, such as that also shown in Fig. 7.1, where RV_1 is used to set an equal mark to space ratio in the output waveform. An alternative approach used in some commercial instruments is simply to use a high-speed analogue switch, operated by a control signal derived from a frequency stable oscillator, to feed one or other of a pair of preset voltages, alternately, to a suitably fast output buffer stage.

An improved Wien bridge oscillator circuit layout of my own, shown in Fig. 7.2 (*Wireless World*, May 1981, pp. 51–53) in which the gain blocks A_1 and A_2 are connected as inverting amplifiers, thereby avoiding 'common mode' distortion, is capable of a THD below 0.003% at 1 kHz with the thermistor controlled amplitude stabilisation layout shown in Fig. 7.2, and about 0.001% when using the improved stabilisation layout, using an LED and a photo-conductive cell, described in the article.

As a general rule the time required (and, since this relates to a number of waveform cycles, it will be frequency dependent) for an amplitude stabilised oscillator of this kind to 'settle' to a constant output voltage, following some disturbance (such as switching on, or alteration to its output frequency setting), will increase as the harmonic distortion level of the circuit is reduced. This characteristic is a nuisance for general purpose use where the THD level is relatively unimportant. In this case an alternative output voltage stabilising circuit, such as the simple back-to-back connected silicon diode peak limiter circuit shown in Fig. 7.3, would be preferable, in spite of its relatively modest (0.5% at 1 kHz) performance in respect of waveform distortion.

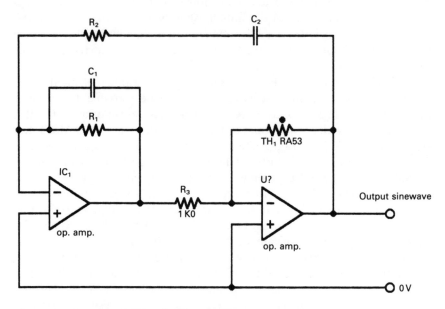

Fig. 7.2 *Improved Wien bridge oscillator.*

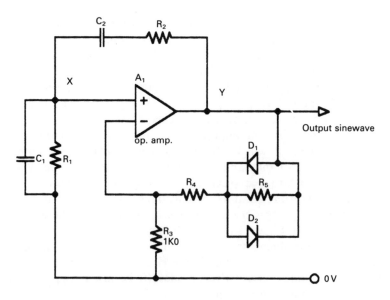

Fig. 7.3 *Diode-stabilised oscillator.*

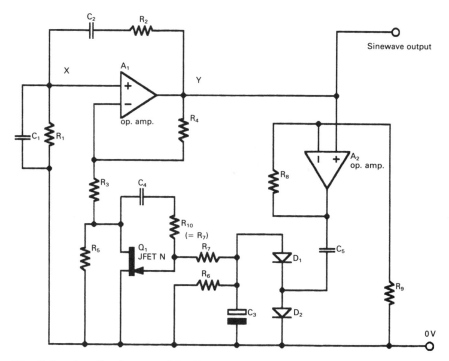

Fig. 7.4 *Amplitude control by FET.*

Rather greater control of the output signal amplitude can be obtained by more elaborate systems, such as the circuit shown in Fig. 7.4. In this circuit, the output sinewave is fed to a high-input impedance rectifier system $(A_2/D_1/D_2)$, and the DC voltage generated by this is applied to the gate of an FET used as a voltage controlled resistor. The values chosen for R_6/R_7 and $C_3/C_4/C_5$ determine the stabilisation time constant and the output signal amplitude is controlled by the ratio of $R_8:R_9$. In operation, the values of R_4 and R_3 are chosen so that the circuit will oscillate continuously with the FET (Q_1) in zero-bias conducting mode. Then, as the $-$ve bias on Q_1 gate increases, as a result of the rectifier action of Q_1/Q_2, the amplitude of oscillation will decrease until an equilibrium output voltage level is reached.

In commercial instruments, a high quality small-power amplifier would normally be interposed between the output of the oscillator circuit and the output take-off point to isolate the oscillator circuit from the load and to increase the output voltage level to, say, 10 V RMS. An output attenuator of the kind shown in Fig. 7.5 would then be added to allow a choice of

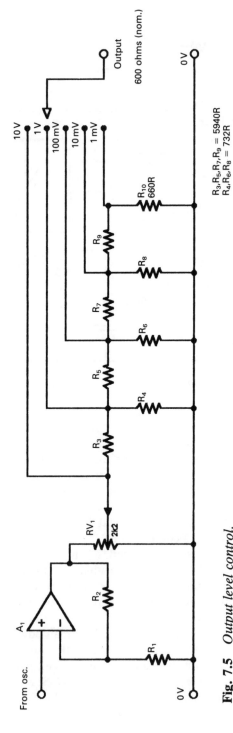

Fig. 7.5 *Output level control.*

maximum output voltage over the range 1 mV to 10 V RMS, at a 600 Ω output impedance.

A somewhat improved performance in respect of THD is given by the 'parallel T' oscillator arrangement shown in Fig. 7.6, and a widely used and well-respected low-distortion oscillator was based upon this type of frequency-determining arrangement. This differs in its method of operation from the typical Wien bridge system in that the network gives zero transmission from input to output at a frequency determined by the values of the resistors R_x, R_y and R_z, and the capacitors C_x, C_y and C_z. If $C_x = C_y = C_z/2 = C$, and $R_x = R_y = 2R_z = R$, the frequency of oscillation will be $1/2\pi CR$, as in the Wien bridge oscillator.

If the parallel T network is connected in the negative feedback path of a high gain amplifier (A_1) oscillation will occur because there is an abrupt shift in the phase of the signal passing through the 'T' network at frequencies close to the null, and this, and the inevitable phase shift in the amplifier (A_1) converts the nominally negative feedback signal derived from the output of the 'T' network into a positive feedback, oscillation-sustaining one.

A problem inherent in the parallel T design is that in order to alter the operating frequency it is necessary to make simultaneous adjustments to either three separate capacitors or three separate resistors. If fixed capacitor values are used, then one of these simultaneously variable resistors is required to have half the value of the other two. Alternatively, if fixed value resistors are used, then one of the three variable capacitors must have a value which is, over its whole adjustment range, twice that of

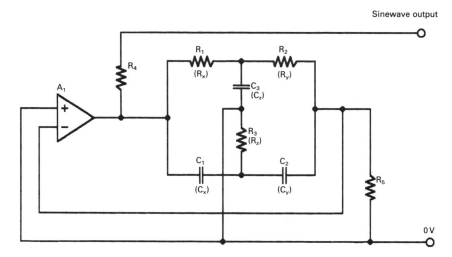

Fig. 7.6 *Oscillator using parallel T.*

the other two. This could be done by connecting two of the 'gangs' in a four-gang capacitor in parallel, though, for normally available values of capacitance for each gang, the resistance values needed for the 'T' network will be in the megohm range. Also, it is necessary that the drive shafts of C_x and C_y shall be isolated from that of C_z.

These difficulties are lessened if the oscillator is only required to operate at a range of fixed 'spot' frequencies, and a further circuit of my own of this kind (*Wireless World*, July 1979, pp. 64–66) is shown in simplified form in Fig. 7.7. The output voltage stabilisation used in this circuit is based on a thermistor/resistor bridge connected across a transistor, Q_1. The phase of the feedback signal derived from this, and fed to A_1, changes from +ve to −ve as the output voltage exceeds some predetermined output voltage level. The THD given by this oscillator approaches 0.0001% at 1 kHz, worsening to about 0.0003% at the extremes of its 100 Hz to 10 kHz operating frequency range.

It is expected in modern wide-range low-distortion test bench oscillators that they will offer a high degree of both frequency and amplitude stability.

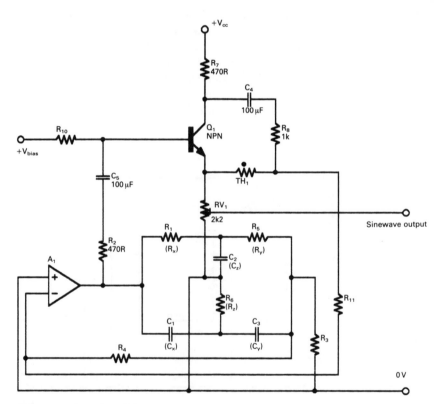

Fig. 7.7 *Level stabiliser circuit.*

This is difficult to obtain using designs based on resistor/capacitor or inductor/capacitor frequency control systems, and this has encouraged the development of designs based on digital waveform synthesis, and other forms of digital signal processing.

Digital waveform generation

Because of the need in a test oscillator for a precise, stable and reproducible output signal frequency a number of circuit arrangements have been designed in which use is made of the frequency drift-free output obtainable from a quartz crystal oscillator. Since this will normally provide only a single spot-frequency output, some arrangement is needed to derive a variable frequency signal from this fixed frequency reference source.

One common technique makes use of the 'phase locked loop' (PLL) layout shown in Fig. 7.8. In this, the outputs from a highly stable quartz crystal 'clock' oscillator and from a variable-frequency 'voltage controlled oscillator' (VCO) are taken to a 'phase sensitive detector' (PSD) – a device whose output consists of the 'sum' and 'difference' frequencies of the two input signals. If the sum frequency is removed by filtration, and if the two input signals should happen to be at the same frequency, the difference frequency will be zero, and the PSD output voltage will be a DC potential whose sign is determined by the relative phase angle between the two input signals.

If this output voltage is amplified (having been filtered to remove the unwanted 'sum' frequencies), and then fed as a DC control voltage to the VCO (a device whose output frequency is determined by the voltage applied to it), then, providing that the initial operating frequencies of the clock and the VCO are within the frequency 'capture' range determined by the loop low-pass filter, the action of the circuit will be to force the VCO into frequency synchronism (but phase quadrature) with the clock signal: a condition usually called 'lock'. Now if, as in Fig. 7.8, the clock and VCO signals are passed through frequency divider stages, having values of $\div M$ and $\div N$ respectively, when the loop is in lock the output frequency of the VCO will be $F_{out} = F_{ck} (N/M)$. If the clock frequency is sufficiently high, appropriate values of M and N can be found to allow the generation of virtually any desired VCO frequency. In an audio band oscillator, since the VCO will probably be a 'varicap' controlled LC oscillator, operating in the MHz range, the output signal will normally be obtained from a further variable ratio frequency divider, as shown in Fig. 7.8. For the convenience of the user, once the required output frequency is keyed in, the actual division ratios required to generate the chosen output frequency will be determined by a microprocessor from ROM-based look-up tables, and the output signal frequency will be displayed as a numerical read-out.

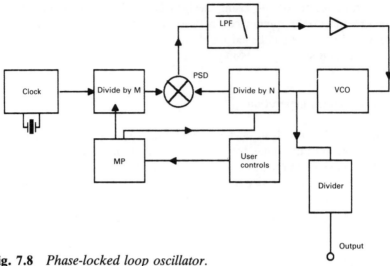

Fig. 7.8 *Phase-locked loop oscillator.*

Given the availability of a stable, controllable frequency input signal, the generation of a low-distortion sinewave can, again, be done in many ways. For example, the circuit arrangement shown in Fig. 7.9 is quoted by Horowitz and Hill (*The Art of Electronics*, 2nd edition, page 667). In this a logic voltage level step is clocked through a parallel output shift register connected to a group of resistors whose outputs are summed by an amplifier (A_1). The output is a continuous waveform, of staircase type character, at a frequency of $F_{ck}/16$.

If the values of the resistors R_1–R_7 are chosen correctly the output will approximate to a sinewave, the lowest of whose harmonic distortion components is the 15th, at −24 dB. This distortion can be further reduced by low-pass filtering the output waveform. A more precise waveform, having smaller, higher frequency staircase steps, could be obtained by connecting two or more such shift registers in series, with appropriate values of loading resistors.

Like all digitally synthesised systems, this circuit will have an output frequency stability which is as good as that of the clock oscillator, which will be crystal controlled. Also the output frequency can be numerically displayed, and there will be no amplitude 'bounce' on switch on, or on changing frequency.

A more elegant digitally synthesised sinewave generator is shown in Fig. 7.10. In this the quantised values of a digitally encoded sine waveform are drawn from a data source, which could be a numerical algorithm, of the kind used, for example, in a 'scientific' calculator, but, more conveniently, would be a ROM-based 'look-up' table. These are then clocked through a

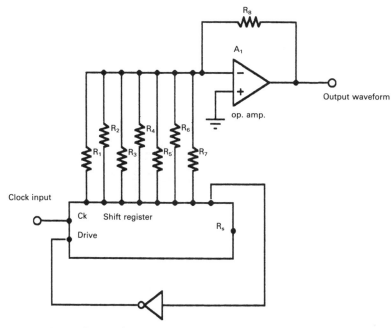

Fig. 7.9 *Digital waveform generation.*

shift register into a 16-bit digital-to-analogue converter (DAC). If the data chosen are those for a 16-bit encoded sinusoidal waveform, the typical intrinsic purity of the output signal will be of the order of 0.0007%, improved by the use of low-pass, sample-and-hold filtering.

Moreover, if digital filtering is used, prior to the DAC, this can be made to track the frequency of the output signal. As in the previous design (Fig. 7.9) the output signal frequency is related to, and controlled by, the clock frequency.

ALTERNATIVE WAVEFORM TYPES

A range of waveforms including square- and rectangular-wave shapes as well as triangular and 'ramp' type outputs are typically provided by a 'function generator'. The outputs from this kind of instrument will usually also include a sinusoidal waveform output having a wide frequency range but only a modest degree of linearity.

An IC which allows the provision of all these output waveform types is the ICL '8038' and its homologues, for which the recommended circuit layout is shown in Fig. 7.11. The output from this is free from amplitude

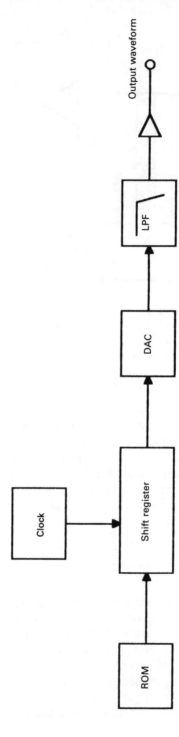

Fig. 7.10 *ROM-based waveform synthesis.*

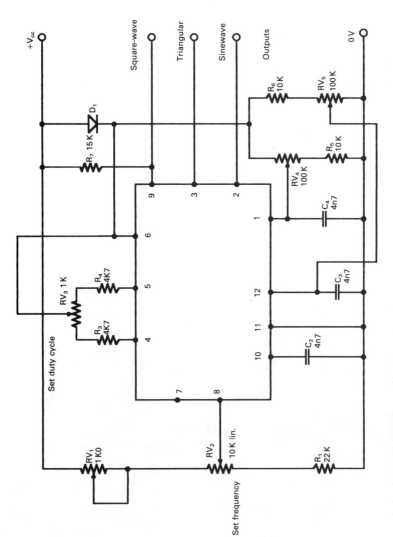

Fig. 7.11 *ICL 8038 application circuit.*

'bounce' on frequency switching or adjustment, and can be set to give a 1 kHz distortion figure of about 0.5% by adjustment of the twin-gang potentiometers RV_4 and RV_5. As shown, the frequency coverage, by a single control (RV_2), is from 20 Hz to 20 kHz.

Since a square-wave signal contains a very wide range of odd-order harmonics of its fundamental frequency, a good quality signal of this kind, with fast leading edge (rise) and trailing edge (fall) times, and negligible overshoot or 'ripple', allows the audio systems engineer to make a rapid assessment both of the load stability of an amplifier and of the frequency response of a complete audio system.

With an input such as that shown in Fig. 7.12(a), an output waveform of the type in 7.12(b) would indicate a relatively poor overall stability, by comparison with that shown in 7.12(c), which would merely indicate some loss of high frequency. The type of waveform shown in 7.12(d) would imply that the system was only conditionally stable and unlikely to be satisfactory in use.

The type of oscilloscope waveform shown in Fig. 7.12(e) would indicate a fall-off in gain at low frequencies, whereas that of Fig. 7.12(f) would show an increase in gain at LF. Similarly, the response shown in Fig. 7.12(g) would be due to an excessive HF gain, while that of Fig. 7.12(h) would indicate a fall-off in HF gain. With experience, the engineer is likely

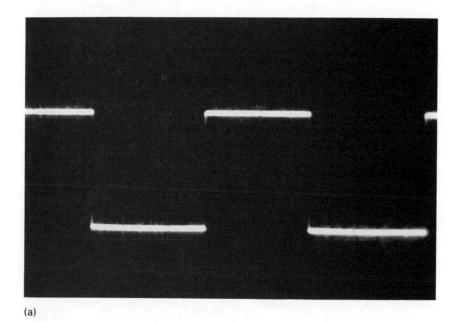

(a)

Fig. 7.12 *(a) Square-wave input waveform (wide bandwidth).*

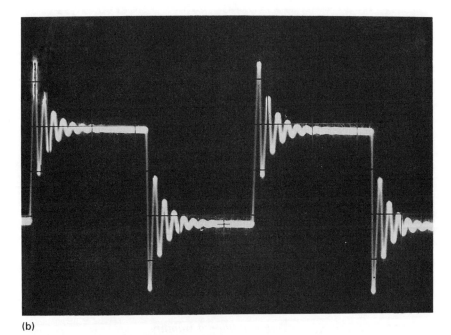

(b)

Fig. 7.12 *(b) poor amplifier stability.*

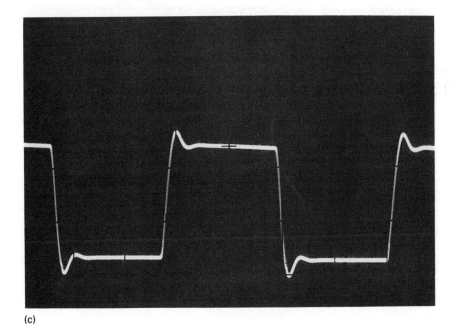

(c)

Fig. 7.12 *(c) relatively rapid (− 12 dB/octave) loss of HF gain.*

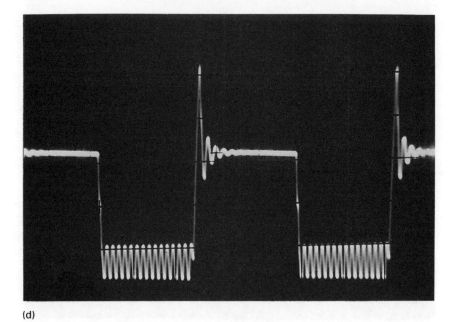

(d)

Fig. 7.12   *(d) output showing conditional stability.*

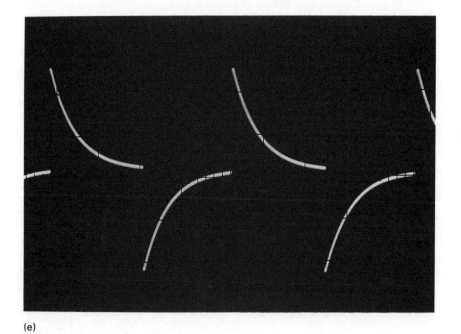

(e)

Fig. 7.12   *(e) poor LF gain.*

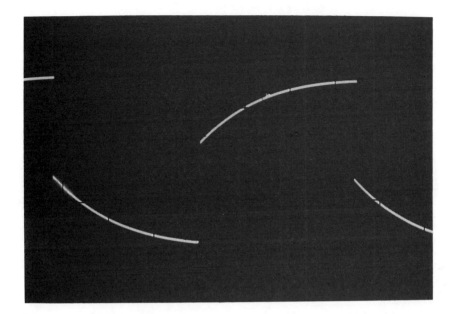

(f)

Fig. 7.12   *(f) increased gain at LF.*

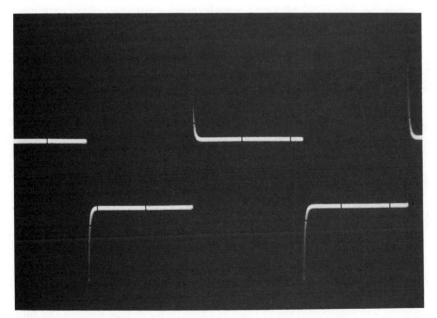

(g)

Fig. 7.12   *(g) excessive HF gain.*

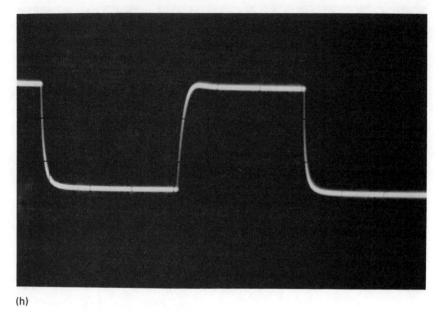

(h)

Fig. 7.12 *(h) loss of HF gain.*

to recognise the types of output waveform which are associated with many of the common gain/frequency characteristics or design or performance problems.

The other types of waveform which can be provided by a function generator have specific applications which will be examined later. The sinewave output, though too poor in linearity to allow amplifier THD measurements to be made with any accuracy, is usually quite free from amplitude variation, and the 'single knob' wide-range frequency control of such an instrument allows fast checking of the performance of such circuits as low- or high-pass filters or tone controls.

DISTORTION MEASUREMENT

One of the more important characteristics of any audio system is the extent to which the signal waveform is distorted during its passage through the system. Where the input signal (f_{in}) is of a single frequency, and this distortion is due to some non-linearity in the transfer characteristics of the signal handling stages, this non-linearity will generate a series of further spurious signals occurring at frequencies which are multiples of the input frequency (or frequencies). Because these spurious signals have a

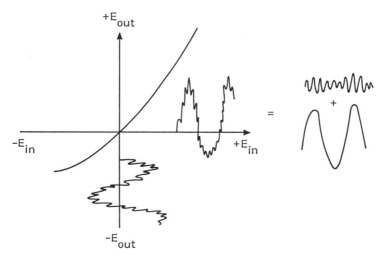

Fig. 7.13 *Origins of IM distortion.*

harmonic relationship (i.e. $2f_{in}$, $3f_{in}$, $4f_{in}$, $5f_{in}$, etc.) to that of the input waveform this characteristic is known as harmonic distortion and, since all such harmonics will be present in the residue, it is usually referred to as 'total harmonic distortion' (THD).

Where the input waveform consists of signals at two or more distinct frequencies, a further type of distortion will occur, in which the amplitudes of each of the signals will be modulated to some extent at the frequency of the others – an effect which is more easily seen if there are only two input signals and these are widely different in both amplitude and frequency, as shown in Fig. 7.13. This type of distortion is known as 'Intermodulation Distortion' (IMD) and leads to a loss of clarity in the reproduction of such complex signals. Since, in audio applications, the input signal will be typically of multiple frequency form, IMD is very important as a design quality. There is, unfortunately, no direct numerical relationship between IMD and THD, except that the lower the THD of a system, the lower the IMD will also be.

THD meters

Two quite distinct techniques are used for such distortion measurement: 'Total Harmonic Distortion' (THD) meters and 'Spectrum Analysers'. In the THD meter, the instrument is usually arranged so that there will be a sharp notch at some point in its gain/frequency curve, which will give zero transmission at some specific frequency, as shown in Fig. 7.14. If this notch

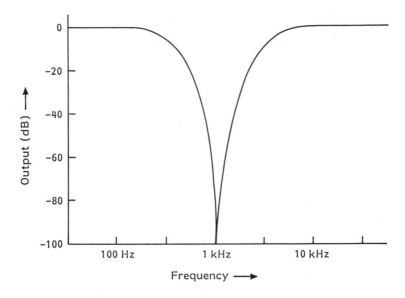

Fig. 7.14 *Parallel T notch filter characteristics.*

is tuned to the same frequency as a low-distortion sinewave input test signal, then, when the test signal is 'notched out', what is left will be the spurious harmonics (plus hum and noise) introduced by the equipment under test, or the test instrument or the associated wiring connections.

In the normal method of use of this instrument, the 100% setting is found if the notch filter circuit is temporarily by-passed, its output is taken to a sensitive wide-bandwidth AC millivoltmeter, and the system gain is adjusted so that the meter reads full scale. Then, when the notch is once more switched into circuit, the meter reading will show the residual THD + noise and hum.

In those cases where the intrinsic hum and noise in the system is fairly low, and the input test signal is a very low distortion sinewave – with, ideally, less than 0.01% THD – a simple THD measurement will often be entirely adequate to determine the performance of the equipment, particularly if the residual output signal waveform, following the notch circuit, is displayed on an oscilloscope.

With experience, the engineer can derive a lot of useful information about the circuit performance from this type of display. For example, in Fig. 7.15, the characteristic 'spikes' due to 'crossover distortion' in the output waveform of a relatively poorly designed (or badly adjusted) push–pull transistor amplifier can be seen very clearly in the THD meter output (and would, in practice, be audibly unpleasant), even though the

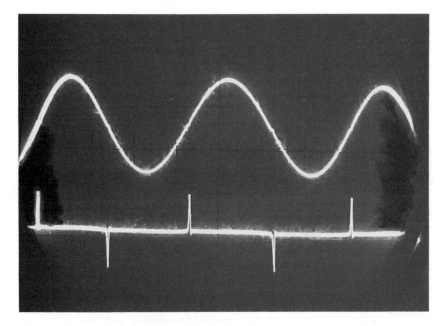

Fig. 7.15 *Crossover distortion residues.*

waveform distortion which caused the spikes cannot be seen very easily in the output waveform shown above.

In Fig. 7.16 the THD meter output waveform shown is that from a high-quality high-power transistor amplifier at an output voltage swing, into a water-cooled resistive load, which is just below clipping level. Because of some slight asymmetry in the operation of the amplifier under test – most probably in the output stage – there is a slight flattening of the lower part of the signal waveform, and this causes the slight undulation in the residual THD display which occurs at these negative-going peaks.

There are also some small crossover-type waveform discontinuities just visible in the THD trace, coincident with the mid-point of the voltage swing. The peak-to-peak amplitude of these is about the same level as that of the random background noise of the amplifier and meter. Because of their brief duration, they represent a very high order harmonic content – probably the thirteenth or higher, and would, possibly be supersonic. Also, because of their brief duration, the energy associated with these glitches is very small. As a matter of interest, the actual harmonic distortion which was indicated for the amplifier test shown in Fig. 7.16 was 0.006%.

A much more obvious distortion residue is that shown in Fig. 7.17. Since the THD residue waveform has twice the number of peaks as the input

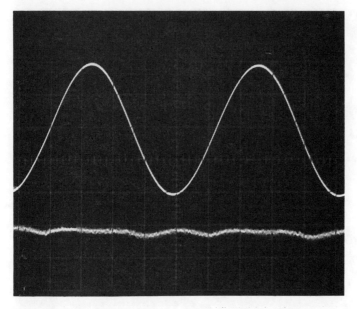

Fig. 7.16   *Distortion residues in high-quality audio amplifier*

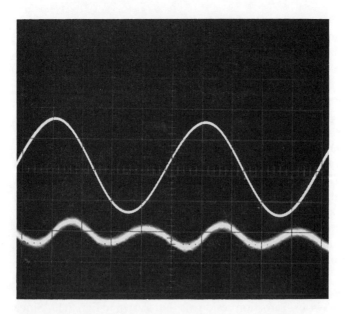

Fig. 7.17   *Distortion from high-quality gramophone PU.*

waveform it is therefore due to the second harmonic. It was, in fact, measured as 0.65%, and was the output from a (high quality) gramophone pick-up replaying a 1 kHz test groove, cut at a lateral velocity of 5 cm/s on an vinyl LP test disc. The noise on the trace is surface noise from the disc. A test signal of this kind is invaluable for setting the pick-up tracing angle, and, as shown, was almost certainly that when the cartridge was optimally adjusted.

As can be seen from the illustrations, a useful feature of this kind of measurement is that it allows the fault in the system to be related to that part of the signal waveform where it occurs. A disadvantage, also shown, is that the presence of high frequency circuit noise, as in Fig. 7.16, tends to conceal the error signals, while that of hum and low frequency noise as shown in Fig. 7.17 (in this case due to surface noise and 'rumble') blurs the display. For these reasons, most typical THD meters include optional low-pass and high-pass filters to lessen unwanted output artefacts. These filters are unsuitable for use at the low- and high-frequency ends of the audio spectrum, and lessen the usefulness of THD measurements under these conditions.

Almost any notch circuit can be used in an instrument of this kind, of which the two most popular are the Wien network nulling circuit, shown in outline form in Fig. 7.18, and the 'parallel T' notch circuit shown in Fig. 7.19. A practical circuit of my own (*Wireless World*, July 1972, pp. 306–308), based on a Wien bridge null network, is shown in Fig. 7.20. In order to get an accurate result the notch must be sharp enough to remove the fundamental frequency entirely, while avoiding any attenuation of the second harmonic frequency. In the circuit shown, the notch is sharpened up by the use of some overall negative feedback between the emitters of Q_4 and Q_1 via R_{13} and R_6.

IMD meters

As noted above, if the transfer characteristics of the system are non-linear, and there are two (or more) signals present at the same time, a further type of distortion will occur, as shown in Fig. 7.13. This is termed 'Intermodulation Distortion' (IMD), and arises because, for any signal, the gain of the system is defined by the slope of the transfer characteristic. This means that, for example, if a large signal is present at the same time as a smaller one, it will have the effect of moving the smaller signal up and down the transfer curve, and its gain, and consequently the size of the output voltage due to it, will change at the frequency of the larger one.

A variety of techniques are used to measure IMD of which the earliest (1961) was that proposed by the (US) Society of Motion Picture and Television Engineers (SMPTE). In this two signals, typically 70 Hz and

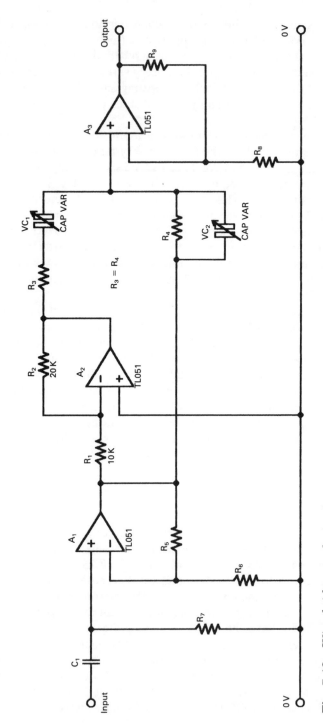

Fig. 7.18 *Wien bridge notch circuit.*

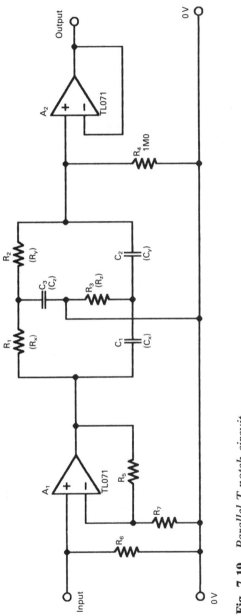

Fig. 7.19 *Parallel T notch circuit.*

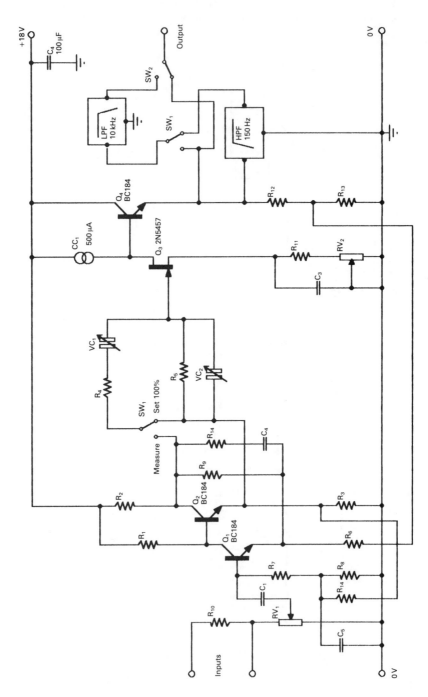

Fig. 7.20 *Wien bridge THD meter.*

7 kHz, at a 10:1 magnitude ratio, are fed to the system, the 7 kHz signal is separated by filtering and then demodulated to recover the 70 Hz intermodulation component which is measured using an RMS reading meter.

An alternative technique which allows the measurement of system performance in the 11–20 kHz region, for which harmonic distortion measurements would not be easily applicable, is the CCIR 'difference frequency' method in which two tones, of equal amplitude, at, for example, 19 kHz and 20 kHz are used as test signals, and the 1 kHz difference tone due to IM effects is isolated by filtering and measured. This type of test can be performed with various input signal frequencies, or even with the input signals swept over the 4–20 kHz range at a constant difference frequency (1 kHz or 70 Hz).

A more recent approach, called the 'Total Difference Frequency' method, proposes test frequencies of 8 kHz and 11.95 kHz. The simple difference frequency product from these will be 3.95 kHz, while the second harmonic of the 8 kHz test tone will generate a 4.05 kHz difference frequency signal. Non-linearities at the HF end of the audio band are shown up by interactions between the third harmonic of the 8 kHz tone and the second harmonic of the 11.95 kHz one, which would produce a 100 Hz difference tone.

These high-frequency tests allow performance evaluations under dynamic conditions as distinct from the more usual single-frequency 'steady-state' tests. This is important, not only because components behave differently under rapid rates of change of operating conditions, and can malfunction in ways not found in steady state, but also because the bulk of the programme material on which audio equipment operates is of a complex electrical form, rich in transients, and the skilled listener can readily detect alien sounds, within this total, due to system errors which may not be revealed by simple steady-state tests.

Spectrum analysis

The most powerful – but also the most complex, the most expensive and the most difficult to interpret – technique for determining system linearity is that known as 'spectrum analysis'. In the audio field, this is done by feeding the system with one or more input test tones, and then measuring the amplitude of all output signal components as a function of their frequency. This will immediately reveal both harmonics of the signal and their intermodulation products, usually shown graphically as amplitude on a logarithmic scale (dBV) as a function of frequency. A good instrument of this kind will allow the display of signal components as low as −120 dB, mainly limited by the noise floor.

Early instruments of this type used a layout, shown in simplified form in Fig. 7.21, which is somewhat similar to that of a radio receiver. In this, the input signal from the system under test, after amplification and buffering by a very low distortion input amplifier, is caused to modulate the amplitude of a swept frequency sinewave signal, typically in the 100 kHz range, which is then amplified by a very narrow pass-band HF amplifier. The output from this amplifier is demodulated and converted into a DC voltage which is fed to a paper chart recorder to give the type of display sketched in Fig. 7.22. The frequency sweep system is coupled both to the chart recorder paper drive and to the variable frequency oscillator so that there will be a precise relationship between the frequency being sampled and the paper chart position. A fundamental problem with this type of instrument is that the response speed of the IF amplifier/demodulator system is directly related to the bandwidth chosen for the IF amp., so, if a high degree of spectral resolution is required (i.e. narrow IF pass-band) the sweep frequency rate and the chart drive speed (which are coupled together) must both be slow, perhaps requiring a minute or more for a complete 20 Hz to 20 kHz sweep.

Modern instruments for use in the audio band employ digital signal processing techniques based on 16 to 20-bit quantisation resolution and sampling rates up to 204.8 kHz, depending on the application. The components of the signal under analysis are then resolved by fast Fourier transform (FFT) techniques. Typical contemporary instruments of this type are the Hewlett Packard 3562A, 35665A and 35670A dynamic signal analysers. Instruments of this type are somewhat more rapid in response than the earlier swept frequency systems, with a typical high-resolution frequency spectrum, depending on the input signal frequency span, requiring less than a minute to display or refresh. A range of outputs would normally be available, including interface data for use with a computer, but if the output required is in the form of a graphical display, this would be provided by a laser or bubble jet printer. A typical numerical analysis of such an encoded FFT signal would allow, as, for example, in the Bruel and Kjaer 2012 instrument, the simultaneous graphical display of individual harmonic distortion components as a function of frequency – a facility which is very useful for loudspeaker, headphone or microphone design.

OSCILLOSCOPES

The remaining major class of test instrument in the audio engineer's laboratory is the 'cathode ray oscilloscope' – normally just called an 'oscilloscope' or 'scope'. This is an instrument which allows a voltage waveform to be displayed, visually, as a graph of voltage as a function of time. Normally, the horizontal axis of the display – the X-axis – will

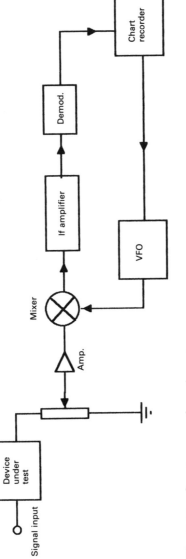

Fig. 7.21 *Spectrum analyser layout.*

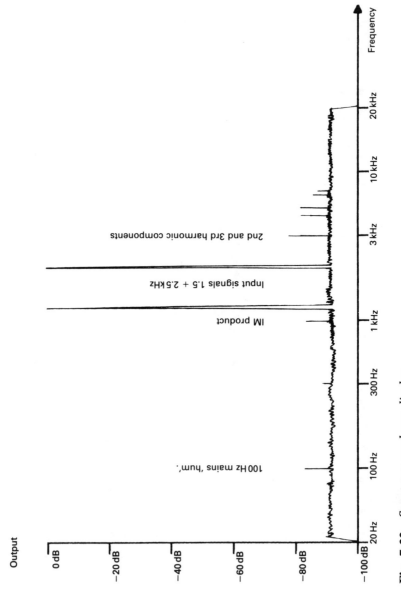

Fig. 7.22 *Spectrum analyser display.*

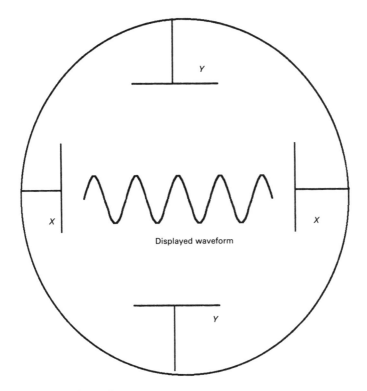

Fig. 7.23 *Typical oscilloscope display.*

represent elapsed time, and the vertical axis – the Y-axis – will represent
the signal voltage being displayed. The way by which this is done is shown,
in a simplified manner, in Figs 7.23 and 7.24.

The oscilloscope tube has an electron source at one end, typically an
indirectly heated thermionic 'cathode', from which a stream of electrons is
drawn and accelerated towards a layer of fluorescent material, the 'screen',
deposited on the inner face of the glass envelope of the tube. The internal
electrodes in the tube cause the electron beam to be focused so that its
region of impact on the screen generates a small, bright, point of light. If
the target material is zinc sulphide, which is an efficient phosphor material,
the emitted light will be of a green colour. Other coloured screens are
available, but green is preferred by most engineers as a restful colour for
extended viewing. A group of four deflection plates is interposed between
the electron source and the screen as shown in Fig. 7.23, and the lateral or
vertical position of impact of the electron beam on the screen will be
controlled by the voltages present on these plates at the instant when the
individual electrons forming the stream pass between them.

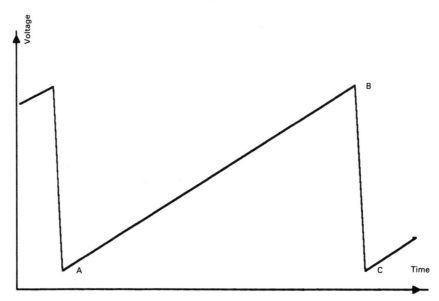

Fig. 7.24 *Timebase scan waveform.*

In a simple 'scope, the horizontal scan voltage will be a 'sawtooth' waveform of the kind shown in Fig. 7.24. In principle, this could be generated by a 'function generator' IC, but in practice, since it is essential that the voltage waveform is extremely linear between points A and B, and occupy as little time as practicable in returning from point B to point C to begin the cycle again, 'scope manufacturers will normally provide their own discrete component circuitry for this purpose, though there will usually be an external electrical input point to allow the user to supply his or her own horizontal scan waveform, when needed. (Since it is desirable to avoid confusion of the display due to the presence of extra images during the scan return from B to C, the trace is usually blanked out during this period.) The voltage waveform which it is desired to display is then applied, after suitable amplification, to the Y plates, which causes the electron beam to be deflected vertically, to an extent which is accurately related to the applied voltage.

The vertical or horizontal position of the display on the tube screen is controlled by the application of a suitable DC voltage to the plates, normally called the 'X-shift' or 'Y-shift' controls. To preserve the sharpness of the spot focused on the screen the electronic circuitry of the X and Y amplifiers is chosen so that the voltages applied to the deflection plates are in 'push–pull' (i.e. equal in magnitude but opposite in polarity). To allow some fine trimming of the focus and deflection voltages, for optimum spot sharpness, an 'astigmatism' control will usually be provided.

The length of time which is occupied in traversing from A to B will determine the time axis of the display, so that, for example, if the waveform displayed on Fig. 7.23 as a result of an alternating voltage applied to the Y plates is a 1 kHz sinewave (i.e. 1 ms/cycle), then the horizontal (X-axis) time scale will be 5 ms, which would imply a 200 Hz repetition rate for the X waveform. The ability of the 'scope to display very high-frequency input waveforms is limited only by the ability of the 'scope electronics to charge and discharge the stray capacitances associated with the plate electrodes, through the inevitable inductances due to the plate connecting leads.

Generally, the 'scope time base generator circuit will allow the user to choose the X-axis repetition frequency, as a switchable sequence of scan times, ranging, perhaps, from 10 s/cm to 0.1 μs/cm. There will also be a control which will allow some variation in these pre-set speed settings to facilitate synchronisation of the signal being displayed with the horizontal scan speed. On some 'scopes, there will also be an option which will enable the user to increase the scan width by a factor of 10. Since the repetition rate remains the same, the effect of this will be to leave the waveform synchronised with the horizontal scan, but to spread this out so that the time component of the waveform can be seen in greater detail, since the X-shift control will allow the waveform displayed to be moved horizontally. The facility of expanding a small portion of the X trace, to speed up part of the display, while leaving the remainder unaffected, is still sometimes offered, as has been the ability to display different traces at different scan speeds, but these are unusual facilities.

An interesting curiosity in this field is that the spot of light on a 'scope screen is able to move at a velocity which is greater than the speed of light – because it is not itself a physical entity, merely a historical record.

The X and Y amplifiers, which amplify the time base and Y input signals before presentation to the deflection plates, are generally similar in sensitivity and bandwidth, to facilitate the use of either of these inputs for signal purposes. A normal maximum Y input sensitivity will be in the range 1–10 mV/cm, with a bandwidth of up to 100 MHz, depending on the cost and performance bracket of the instrument. For audio use a 2 mV/cm sensitivity and a 10–20 MHz bandwidth would be quite adequate. The need for such a high HF bandwidth arises because, particularly with MOSFET devices, circuit instabilities may lead to spurious oscillation in the 5–20 MHz range, and it is desirable that the 'scope should be able to show this, when it is present.

It is also normally expected that the Y amplifier should be 'direct coupled', so that a DC voltage can be displayed as a vertical offset of the horizontal scan position. The 'shift' voltages are added to the Y (or X-axis) signals, after the input attenuation, so that the required scan position can be obtained under signal input conditions where both AC and DC inputs are present. If it is preferred to ignore the DC component of the signal, an input blocking capacitor can be switched into circuit. The 'direct coupled' input facility is very useful for measuring small DC offset voltages in sensitive regions of the circuitry.

Multiple trace 'scope displays

With the exception of small, portable or battery operated instruments, almost all modern 'scopes offer multiple 'trace' displays, which enable two or more Y-axis signals to be shown simultaneously, one above the other on the screen. This facility is exceedingly valuable, in that it allows the comparison of the input and output waveforms from any system under test, so that changes in waveform shape, or delays in time between input and output, can be seen immediately.

When using a simple notch-type THD meter, a dual trace display is also very helpful in that the input waveform can be shown on, say, the upper trace, and used for display synchronisation, while the noise and distortion residues are displayed on the other. Not only does this allow the identification of the position on the waveform at which the defect occurred, but it also prevents the loss of synchronisation which would otherwise take place when the input waveform was 'notched out'. With systems using digital signals multiple trace displays allow the timing of a number of output or input pulse streams to be checked, though this is an area somewhat outside the scope of this book.

In early 'scopes, the provision of multiple trace displays was achieved by the use of separate electron 'guns', focusing electrodes and X and Y plate assemblies, but this made for bulky and expensive CR tubes. In modern 'scopes, the technique employed uses a combination of fast electronic switching applied to the input signal in synchronism with a superimposed square wave vertical deflection waveform of adjustable size. This allows one or other of the input signals to be switched to the Y amplifier, and thence to the Y deflection plates, at a vertical display position determined by users by that choice of square wave size. The display sequence may be either 'alternate', in which sequential 'X-axis' sweeps are chosen to display alternate inputs, or 'chopped', in which the input waveforms are rapidly sampled, during the sweep, and routed, up or down, to the chosen display positions.

Normally the choice of trace division technique is determined automatically by the 'scope circuitry according to the X-axis sweep speed selected. At low sweep speeds, the input waveforms will be chopped at a frequency which is high in relation to the signal frequency. This would be preferable, for example, if it was desired to do time comparisons between waveforms from different Y inputs. For high-frequency signals, alternate displays are normally selected, on the grounds that the persistence of vision will make the display seem continuous. The accuracy of the relative horizontal position of such a display is, in this case, dependent on the freedom from 'jitter' of the time-base synchronisation.

Time-base synchronisation

The ability to 'freeze' a repetitive waveform on the 'scope display is an essential quality in any oscilloscope, and facilities are invariably provided to allow trace synchronisation with any specified signal input, but also from an external 'sync' input. Various types of synchronisation are also usually provided by switch selection, in the hope that some of these may be more effective that others. My choice of words is prompted by the experience, over many years and with many 'scopes, that some 'scopes and some waveforms can prove very difficult to lock on the screen. In fact, one of my personal priorities in the choice of an oscilloscope is ease of synchronisation. Sadly, one only ever discovers whether this is good or poor after the purchase of the instrument – unless one has had previous experience with that make and model.

External X-axis inputs and Lissajoux figures

The input to the X-axis amplifiers can usually be disconnected from the output of the sawtooth waveform generator and taken from an external 'Time Base Input' connection. This then allows a single-beam 'scope to be used to display 'Lissajoux figures' of the kind shown in Fig. 7.25. In the ring-shaped display of Fig. 7.25(a), the external X input and Y inputs are at the same frequency, though at phase quadrature. If they are in phase, or in phase opposition, the resultant display will be a straight line at 45° or 135° to the horizontal axis.

In Fig. 7.25(b), the X input is at twice the frequency of the Y one, while in Fig. 7.25(c) the Y input is twice the frequency of the X input. In Fig. 7.25(d) the X input is three times the frequency of the Y input – and so on. This provides a rough and ready means of assessing the frequency of an input sinusoidal signal if one has a calibrated sinewave source. However, small frequency drifts in the signal inputs lead to continuous snake-like intertwinings of the display, which makes the precise determination of frequency difficult to achieve.

With a dual trace oscilloscope, the same result may be obtained more easily by displaying the signal of unknown frequency, in synchronism on one trace, and then using the second trace to display the output from a variable, but calibrated signal generator. Apart from the difficulty of

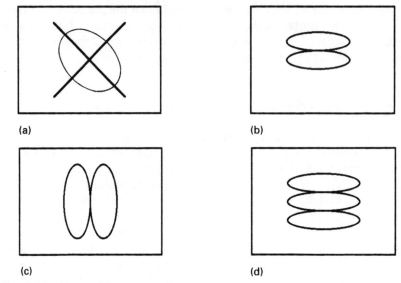

Fig. 7.25   *Typical Lissajoux figures.*

obtaining simultaneous synchronisation, the unknown frequency can be determined from noting when both traces display the same number of cycles.

An interesting application of the Lissajoux principle is shown in Fig. 7.26, where the LH and RH channels of a stereo signal, such as, for example, an FM tuner, are connected to the X and Y inputs of the 'scope. A mono signal will give the kind of display shown in Fig. 7.26(a), where the length of the line display will depend on the magnitude of the input signal.

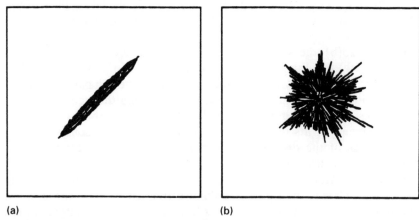

Fig. 7.26   *Mono/stereo signal display.*

On a speech or music input, a stereo signal will give the kind of display shown in Fig. 7.26(b), where the extent of the 'roundness' of the pattern is an indication of the extent of the image separation of the two channels. This allows the adjustment for maximum channel separation of an FM stereo decoder, for example, without the need to seek special test signals.

'X-axis' signal outputs

Most 'scopes also provide a 'time-base output' socket, at which the '*X*-axis' scan voltage waveform (Fig. 7.24) is available. This is required for controlling the frequency sweep of the voltage controlled RF oscillator in a frequency modulated oscillator, or 'wobbulator', whose use was described in Chapter 2 (Tuners and Radio Receivers), and for which the output waveforms were shown in Figs 2.41–43. A similar system is used for the 'Panoramic' display of radio signals as a function of frequency.

Digital storage oscilloscopes

Oscilloscopes, like all other electronic test equipment, has become more versatile, more complicated and more expensive with the passage of time, and this development is particularly notable in the field of 'storage' oscilloscopes, which allow the provision of numerically stored displays. The ability to 'freeze' a transitory waveform, so that it could be measured or examined at leisure, has been sought for very many years, and early instruments endeavoured to meet this need by the use of long-persistence screen phosphors, which would continue to display the screen waveform, though at a much reduced brilliance, for a minute or more following switch-off. An improvement on the performance of long-persistence screens was obtained by the use of a dual-phosphor screen coating, in which a normally dormant phosphor layer could be reactivated to display its last image, by flooding it with infra-red radiation. However, contemporary digital technology, and fast A/D converters, offer the ability to capture and store the actual signal in numerical form – information which is refreshed each scan. In the technique used, the incoming signal is repetitively sampled, at a rate which is high in relation to the horizontal sweep speed, and the instantaneous signal voltage level is stored for a time which is long enough for its value to be noted and transformed (quantised) into a digitally encoded form.

This instantaneous numerical equivalent to the input signal amplitude is then stored in a fast random access memory store, while the output from a digital-to-analogue converter (DAC) is simultaneously displayed on the oscilloscope screen. The advantage of this type of display – apart from the convenience of being able to examine a possibly transient waveform at one's leisure – is that, once the signal is in digital form, a wide range of mathematical manipulations may be performed on it. An example of this is the Fast Fourier Transform process noted above in relation to spectrum analysis. The stored signal can also be fed to a computer or printer for an immediate hard copy.

It is also common, with modern digital storage oscilloscopes, to use the microprocessor system within the control electronics to show the X-axis time scale on the 'scope display, as well as the DC and AC parameters, including frequency, of the input signal(s). Input signal sampling rates of up to 500 Mega-samples/s are available in top-range instruments – though these are expensive. Vertical resolution equivalent to '8-bit' encoding (±0.15 mm on a 100×125 mm tube face size) is typical.

Modern test instrumentation of the various kinds described above allows a great deal of information to be obtained about the electrical characteristics of audio equipment. However, as indicated at the beginning of this chapter, our understanding of the relationship between the measured performance and the quality of sound produced by the equipment under test is still less comprehensive that we would wish, and innovations in technology, particularly in the field of digital audio, continually pose new questions about the acoustic significance of those imperfections which remain.

CHAPTER 8

Loudspeaker crossover systems

WHY NECESSARY?

In the majority of cases, the output from an audio system will be fed to some form of electro-acoustic transducer, such as a loudspeaker (LS) or headphone, whose purpose is to convert the electrical input signal into a sound pressure wave which will evoke a response in the ear of the listener. Ideally, this transducer will have a constant acoustic output, for a given electrical input signal, over the range 20 Hz to 20 kHz, or whatever the human ear is supposed to be able to hear.

In practice, there is no known LS driver system (I am using this term in the sense of the individual moving-coil, magnet and cone assembly, or its equivalent, rather than the complete unit in wooden box, which I will call a 'loudspeaker') which is capable of operating completely satisfactorily over the whole of this frequency span. The loudspeaker manufacturers seek to approach this ideal by combining a number of individual units with suitable 'crossover' networks, so that their combined outputs will achieve the desired result. The units used in such a multiple driver system will mostly be based on moving coil (m/c) layouts but the same limitations will apply, though in a somewhat different sense, to electrostatic or ribbon LS systems.

In general, with an m/c driver unit the optimum frequency span, in which it will be free from major peaks and troughs in its acoustic output, will cover a range of about ten or perhaps twenty to one, of which the lower frequency end will be determined, usually, by the natural (bass) resonance of the cone assembly. This makes it (just) possible to cover the range from 50 Hz to 20 kHz with a two-speaker system, though it may be done more easily, and with a greater bass extension, with a three-speaker assembly. The reasons for the inability of the m/c driver unit to provide a constant acoustic output over a wide frequency range are complex, but, as a simplified analysis, can be considered as a result of the poor mechanical coupling between the moving coil and the cone, and the failure of the LS cone itself, when urged by the coil, to move as a rigid piston to displace enough air to produce the required acoustic output.

I will consider these 'first-order' problems separately, but the ways in which adequate – though not perfect – solutions are found to these owes as much to art and intuition as it does to science or engineering. This is why there are so many good but different LS designs on sale, and why LS designers continue to offer new designs.

CONE DESIGN

Effects of cone flexure and cone break-up

In its simplest form, an LS diaphragm is simply a rigid piston, mounted for operational purposes in an opening, ideally of the same size as the piston, in an infinitely extended sound-proof wall (acting as a 'baffle' to prevent the displaced air from simply flowing round the edges of the piston from front to rear) as shown in Fig. 8.1(a). When the piston is caused to move, it will displace a quantity of air which is proportional to its axial movement multiplied by its surface area.

Unfortunately, because no such piston is completely rigid, the part of the piston to which the actuating mechanism (the coil and magnet) is attached will move more than the periphery of the piston and this will reduce its

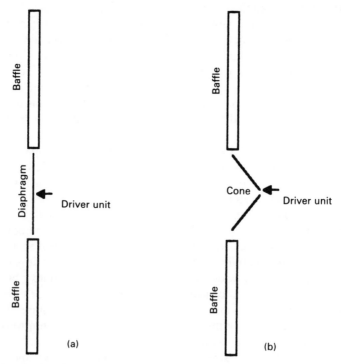

Fig. 8.1   *Basic LS action.*

efficiency, especially at higher frequencies, where the effect of the air loading is greater. Also, the differential movement of the centre of the piston in relation to its periphery due to diaphragm flexure implies that there will be surface wave motion along the piston. This will complicate the acoustic radiation characteristics of the system, and this difficulty will be compounded by the need for some kind of flexible surround to bridge the gap between the edge of the piston and the aperture in which it is mounted. This surround will cause an acoustic mismatch, and will result in any surface waves being reflected back along the surface of the piston, to disturb yet further the sound radiation characteristics.

If the diameter of the piston is reduced, this will increase its effective stiffness, but its axial movement will need to be proportionately increased if it is to be able to displace the same volume of air. Some designers claim that this can cause problems due to Doppler effects (frequency shift through movement of the sound source) if the same piston is required to handle signals at different frequencies, simultaneously – especially if one of these signals is at a low frequency, which requires a larger piston movement. For these reasons, flat LS diaphragms, which will not be very stiff, are seldom found, though some composite structures, based on sandwiches of metal foil and rigid polystyrene foam, have been used with success as 'woofers' required to cover only one or two octaves at the bottom end of the audio band.

Although electrical 'crossover' systems will reduce this type of problem, by limiting the frequency coverage required from any individual LS driver unit, the normal design approach is to make the piston conical in form, as shown in Fig. 8.1(b), rather than flat. This results in a very great increase in stiffness for a given diameter and thickness of the cone material. However, while the behaviour of the cone is good at the low-frequency end of its usable frequency range, as the frequency is increased the cone will cease to behave as a simple piston and the cone surface will tend to break up into various 'ringing' modes, akin to the ringing of a shallow wine glass. These modes will cause unwanted peaks and troughs in the radiated frequency response.

The most common choices of the LS designer for cone materials are soft, felted paper; often with some circumferential corrugations to encourage the inner sections to break away from the outer sections of the cone at high frequencies, and act as smaller, more rigid, pistons; or cones moulded from some inert, heavily internally damped, thermo-plastic, on the grounds that this will not be excited easily into any structural resonances. Alternatively, the cone may be moulded in a somewhat curved or exponential cross-section, because this is not subject to bell-type resonances. However, although this may extend the frequency span which can be covered by any individual LS driver unit, all these solutions result in lower transducer efficiency, which is undesirable.

Rigid cone systems

An alternative approach which has been exploited from time to time, most notably in the UK by Jordan–Watts Ltd and by the GEC, is the use of cone materials which have a very high stiffness-to-weight ratio, such as aluminium or titanium, and then to limit the diameter so that the very prominent HF resonances, which will inevitably occur in such systems, are high enough in frequency to be effectively outside the required audio spectrum. The piston area required to allow useful acoustic outputs at low frequencies is then achieved by connecting a number of these drive units so that their diaphragms move in unison. The crossover network in this case is a load-sharing system rather than, as is more usually the requirement, a frequency band separation arrangement.

Coil attachment and magnet characteristics

For a high LS efficiency, for any given cone structure, the magnetic flux in the air gap in which the coil moves should be as high as possible. This is attained, in part, by the use of very high energy magnet systems, based on exotic alloys, and in part by keeping the air gap between the pole pieces, in which the coil moves, as narrow as possible. However, this, in turn, leads to problems because it is then necessary to provide some centering mechanism attached to both the coil and the LS frame, to make sure that the coil is held so that it cannot touch the magnet pole pieces. This inner centering provides the bulk of the restoring force applied to the cone, and it cannot be made too stiff or it will increase the fundamental resonant frequency of the LS to an undesirable extent.

The fundamental resonant frequency of the cone assembly is of importance because the output will fall at −12 dB/octave below this frequency – though cabinet design can arrest this fall somewhat. The sound waveform of signals below this frequency will become increasingly distorted by the presence of second and third harmonics of the input signal.

A further aspect of coil construction is that, in high-power units, considerable power may be dissipated in the coil, and the heat must be either conducted or radiated away. This can be facilitated by winding the coil on an aluminium former, which could be bonded to a short aluminium cone, fixed to the centre of the LS diaphragm. However, this generates a further acoustic mismatch in the cone assembly, which introduces additional output irregularities. An electrically conductive speech coil former is also of use in assisting in damping the resonances of the cone, insofar as these are transmitted to the speech coil.

SOUNDWAVE DISPERSION

Even if it were possible to design an LS driver unit which could cover the whole of the audio frequency band, it would still be necessary for this to have a relatively large cone area to be able to generate an adequate volume of sound at the lower audio frequencies. This leads to the difficulty that where the wavelength of the sound waves becomes small in relation to the width or height of the radiator, any radiating system will concentrate the energy of the sound radiated by it into a relatively narrow beam in the horizontal or vertical sense. This will be unsatisfactory for normal listening, where, for the comfort or convenience of the listeners, it is preferred that the radiated sound shall be dispersed over a wide area.

CROSSOVER SYSTEM DESIGN

In principle, any low-pass or high-pass filter system, or a combination of these, could be used to select the portion of the frequency spectrum which is allocated to the bass, mid-range and HF driver units. (Where they are connected in the LS circuit within the cabinet, they will almost always be based on LC circuit layouts in order to provide the low circuit impedances needed to drive m/c LS units.) Examples of first-, second- and third-order LC filter circuits are shown in Fig. 8.2 but, in practice, with normal m/c driver units, only second-order (-12 dB/octave) filter circuits will be used, of the general kind shown in Fig. 8.2(b). This is because it is essential that the electrical inputs fed to the LS drivers shall be capable of being added together so that there is no significant hump or trough in the combined LS frequency. This is only easily possible with second-order layouts.

For the notional 1 kHz crossover frequency which I have chosen for the transfer of input signal from a low-frequency driver unit, to a mid-range unit, the low-pass (LF driver) input filter (Fig. 8.2(b)) will have an electrical output (shown as G on the graph) of the kind shown in Fig. 8.3. The high-pass (HF driver) input filter will have an output of the kind in Fig. 8.4. If these two outputs are added together acoustically by two equally efficient drive units, their combined output will be as in Fig. 8.5, which shows a -6 dB trough in the gain curve at the crossover frequency, coupled with an abrupt phase shift (as in the curve labelled P) – a situation which would not be acoustically satisfactory.

In the case of the second-order filter, however, though not in the first- or third-order systems, it is possible to reverse the polarity of the signal fed to one or other of the LS drivers. In the case illustrated I have exchanged the lead/terminal connections to the high-frequency unit. This results in the linear phase characteristics of the result shown graphically in Fig. 8.6, where the -6 dB trough is converted into a much more tolerable 0.95 dB

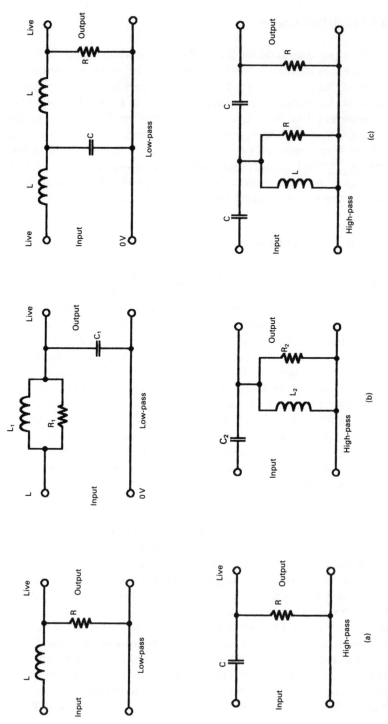

Fig. 8.2 *LC crossover networks.*

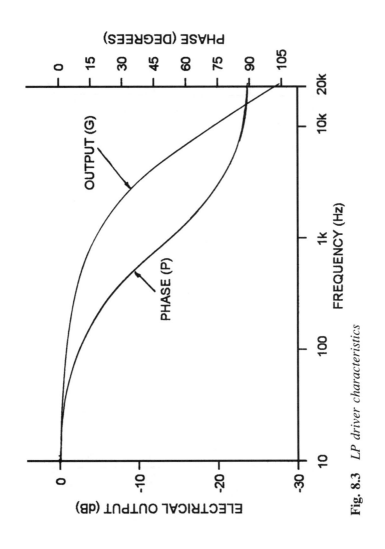

Fig. 8.3 *LP driver characteristics*

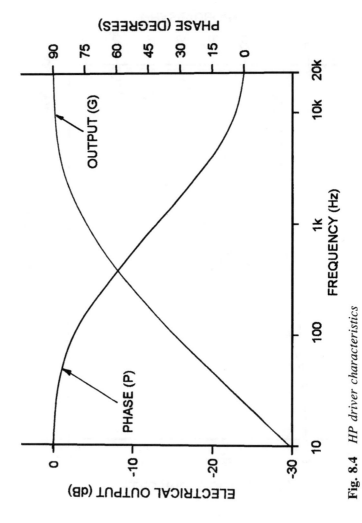

Fig. 8.4 *HP driver characteristics*

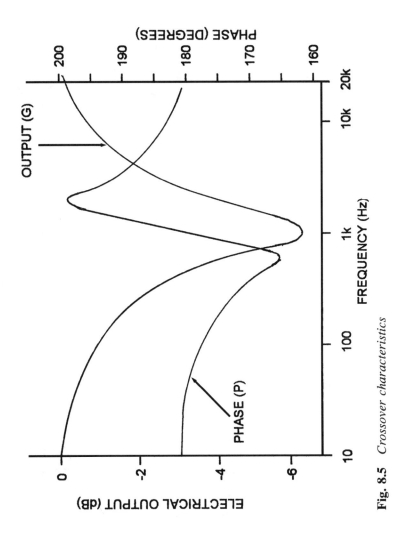

Fig. 8.5 *Crossover characteristics*

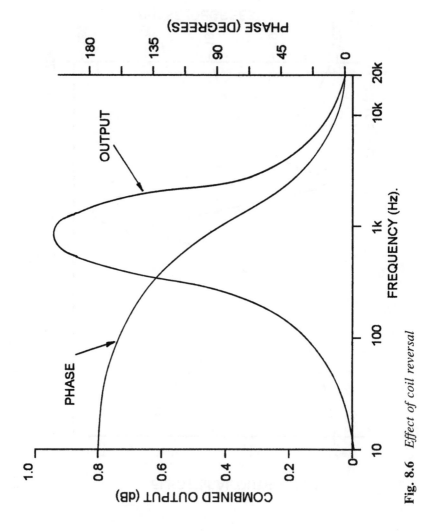

Fig. 8.6 *Effect of coil reversal*

hump in the frequency response. (In practice, the component values of the filter would be adjusted slightly to move the cut-off frequencies slightly further apart to reduce this hump to near zero.)

The manipulation of crossover network systems was considered, at some length, by Pramanik of Bang and Olufsen (S. K. Pramanik, *Wireless World*, November 1975, pp. 529–532). His explorations of the various network possibilities showed that while the second-order filter system was the best of the three listed above, this still left discontinuities either in gain or in phase. He therefore proposed the use of an active crossover network, from which a third output was derived which could be fed to a supplementary driver unit to fill in the troughs resulting from conventional crossover systems.

A problem which will always arise in practice is that the individual driver units will vary in sensitivity not only from one to another of an apparently identical type but also, and more understandably, between driver units of different types, intended to cover different parts of the audio spectrum.

The use of resistors within the crossover networks is to be avoided where possible because they will dissipate energy (and get hot) and will also reduce the system efficiency. Those shown in Fig. 8.2, inserted to improve the theoretical smoothness of the filter slope, would be dispensed with if the characteristics of the driver unit or its enclosure contributed enough damping without them. Therefore the idea of using an output potentiometer to adjust the voltage drive to the LS unit is quite impracticable.

CROSSOVER COMPONENT TYPES

Although in low-cost, modest performance loudspeaker systems it is normal practice to use reversible electrolytic capacitors and iron-cored inductors, in order to reduce size and the cost of providing the relatively large inductance and capacitance values needed, the better-quality systems will invariably use air-cored inductors and non-polar capacitors. The latter should preferably be made from low-loss dielectric materials such as polycarbonate or polypropylene.

If an adjustment of signal input to one unit relative to another is essential this will usually be done by a small coupling transformer, and in high-grade systems this will also need to be an air-cored unit. However, this approach should be avoided if at all possible since it will probably impair the network characteristics.

The design of crossover networks is additionally complicated by the fact that the LS driver unit, within an enclosure, does not present a pure 4, 8 or 15 ohm, frequency-independent, resistance at its input – a condition assumed as a first approach to the design of the crossover system. In practice, in addition to the quite significant electrical resistance and inductance of the speech coil winding, all the mechanical resonances in the LS cone, or its mounting, or its enclosure, will be combined in its input impedance, together with the mechanical components of capacitance, resistance and inductance due to the compliance, frictional losses, and mass of all the moving or vibrating parts of the system.

LS OUTPUT EQUALISATION

As has been seen, most LS driver units do not have a constant acoustic output as a function of frequency for a constant input signal voltage. However, it is possible, with most such systems, to electrically equalise at least partially the response of the driver unit by the insertion of strategically placed inductors or capacitors in series or in parallel with the driver unit. Typical circuit arrangements which will produce a dip or a roll-off in the frequency response of the system are shown in Fig. 8.7. The values of resistors, inductors or capacitors in the circuit will be chosen to

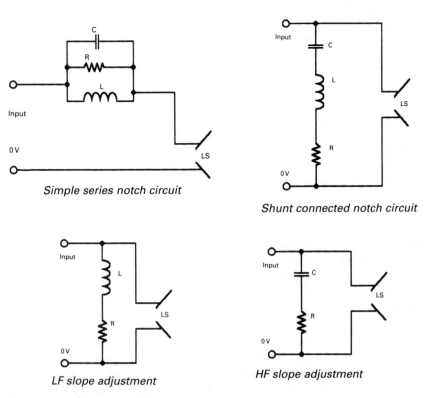

Simple series notch circuit

Shunt connected notch circuit

LF slope adjustment

HF slope adjustment

Fig. 8.7 *Response-equalising circuits.*

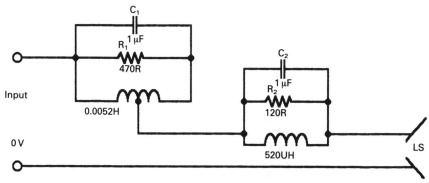

Fig. 8.8 *Baxandall's equalising circuit for ELAC RM/109 LS unit.*

provide the results required, bearing in mind the complex impedance characteristics of the actual driver unit, which will affect the final results.

Circuit layouts of this type are often used by commercial loudspeaker manufacturers to minimise the peaks and troughs which their frequency response measurements have revealed during prototype testing. An example of this approach was described by Baxandall (P. J. Baxandall, *Wireless World*, August 1968, pp. 242–247) with reference to a design which he had built for his own use with a single 15 ohm, 9 in. × 5 in. ELAC RM/109 LS unit, of which the circuit he used is shown in Fig. 8.8. The effectiveness of this approach, in the case of the simple and inexpensive LS unit employed, was attested by the flattering comments subsequently made by several experienced hi-fi reviewers.

ACTIVE CROSSOVER SYSTEMS

The difficulty of making precise inductor/capacitor crossover systems, at LS line impedances, which will nowadays be either 4 or 8 ohms, and the inconvenient load characteristics of the LS unit itself – which are complicated by the motional impedances of the unit and its housing, and the LS coil inductance and resistance – has encouraged the more perfectionist loudspeaker manufacturers (and experimentally minded hi-fi enthusiasts) to transfer the crossover and equalisation networks to the input of the power amplifier, which is a high impedance point. This approach leads to the types of layout shown schematically in Figs 8.9 and 8.10. Here an electronic filter system is interposed between the signal input and the power amplifier, and the output of the power amp is connected directly to the LS driver units. In hi-fi jargon, these arrangements are commonly called 'bi-amplification' or 'tri-amplification', depending on the number of LS units used.

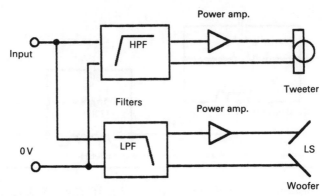

Fig. 8.9 *'Bi-amplifier', layout.*

There are many advantages in this system, of which the principal ones are that the LS coil is driven from a very low source impedance. This leads to:

- improved damping of cone resonances
- the filter and equalisation arrangements can all be implemented by the use of low-power components
- the filter designs can have more nearly ideal phase and amplitude characteristics
- the pass-band handled by the individual amplifiers is reduced, which lessens the exposure of the circuitry to intermodulation distortion effects

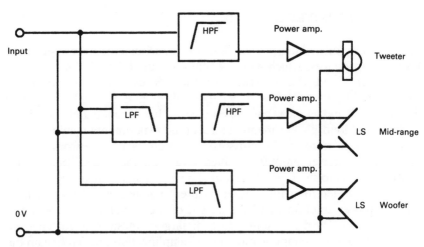

Fig. 8.10 *'Tri-amplifier' system.*

- the output power requirements of the individual amplifiers can be tailored more precisely to the relative energy distribution of the audio signal spectrum
- the relative magnitudes of the signals fed to the driver units can be adjusted, precisely, by trimmer potentiometers without the loss of energy and the dissipation of heat which would otherwise be incurred.

Additionally, the power amplifiers can be housed in the loudspeaker enclosures, which avoids those problems that can occur with external amplifier–LS connector cables, since the only long connections, in the living room of the user, will be between the pre-amplifier and filters, and the power amplifiers inside the loudspeaker units. These will be at high impedance, where any good-quality screened cable will be entirely adequate.

The only major snags with this approach are that it is much more expensive than a conventional system, and that it commits the user to the power amplifier chosen by the loudspeaker manufacturer.

ACTIVE FILTER DESIGN

Although, in principle, any even-order (e.g. second, fourth, sixth, etc.) Butterworth response filter will satisfy the requirement for gain uniformity and phase linearity at the crossover point, the use of higher-order filter types will generally require that some additional phase compensation should be included in the crossover circuit. The simplest, and most popular, arrangement is that in which the filters shown in the schematic layouts of Figs 8.9 and 8.10 are second-order Sallen and Key designs of the kind in Figs 8.11 and 8.12.

In both the high-pass and low-pass circuits shown in Figs 8.11(a) and 8.12, the -3 dB 'turnover' frequency (f_T) will be given by the relationship $f_T = 1/(2\pi\sqrt{(C_1C_2R_1R_2)})$. For example, in Fig. 8.11(a), for $C_1 = C_2 = $ 4n7F, and $R_1 = 18$K and $R_2 = 56$k, $f_T = 1$ kHz. In Fig. 8.12, a 1 kHz turnover frequency would be given by $R_1 = R_2 = 33$k, where $C_1 = 8$n2F and $C_2 = 2$n7F. For the values shown the circuits will have a Q of $1/\sqrt{(2)}$ (0.707). The Q can be increased, giving a somewhat steeper attenuation rate (but also a greater phase shift at f_T) if the gain of IC_2 is increased, as shown in Fig. 8.11(b).

ATC Loudspeaker Technology Ltd, who are one of the foremost European suppliers of professional LS monitor units in which active crossover systems are incorporated within the loudspeaker housings, precede both the high- and low-pass second-order filters with a state-variable filter, having a Q of unity, to generate a fourth-order Butterworth response filter characteristic. They also provide individual

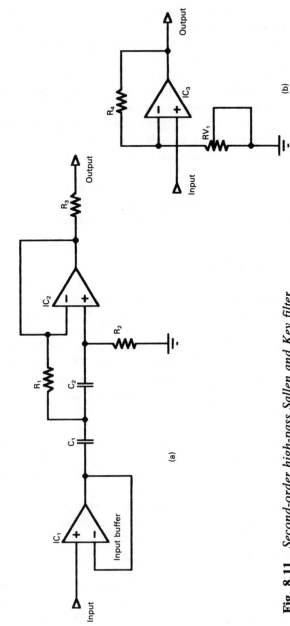

Fig. 8.11 *Second-order high-pass Sallen and Key filter.*

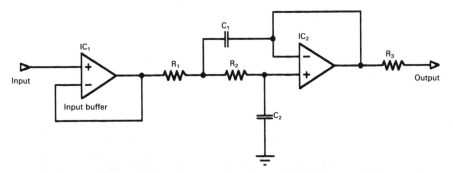

Fig. 8.12 *Second-order low-pass Sallen and Key filter.*

pass-band gain adjustment controls, used to compensate for inequalities in driver unit sensitivities in setting up the system, and a fast-acting 'soft clipping' limiter. This serves both to protect the driver units and also to allow the loudspeakers to be operated at a higher mean sound level without the audibly unpleasant effects which would arise if the power amplifiers were driven into 'hard clipping'.

ATC also incorporate unity-gain phase correcting 'all-pass' filters, of the kind shown in Fig. 8.13, in their crossover divider networks. These allow compensation for the phase errors due to the different sound path lengths from the driver unit to the ear of the listener, and can also remedy any imperfections in the gain/phase characteristics due to acoustic superposition in the crossover region. A photograph of the ATC active crossover system is shown in Fig. 8.14.

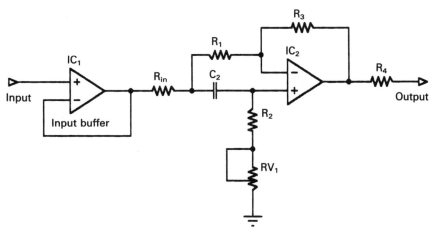

Fig. 8.13 *Phase-correcting filter.*

Fig. 8.14 *ATC active crossover system.*

BI-WIRING AND TRI-WIRING

If the cause of the first major parting of the ways of the 'engineers' and the 'audiophiles' was the audibility of class 'B' type crossover distortion in transistor-operated audio amplifiers (nothing to do with loudspeakers in this case) – which the engineers said was either non-existent or negligible, and the audiophiles said they could hear and disliked – the second was whether or not loudspeaker cables were audibly important. Once again, the engineers said that their calculations and measurements proved conclusively that they were unimportant acoustically, provided that they were of a reasonably low resistance, and, once again, the audiophiles said that they could hear a difference, and they were acoustically very important.

My own feeling in this instance is that the audiophiles have overstated their case but, nevertheless, there is a grain of truth in their claims, as can be shown by a simple experiment. Let us take the case of a pair of LS units which have been wired up using a couple of equal lengths of twin-core 10 amp mains cable. This is a fairly common practice among the more prosaic, or more frugally minded, members of the audio fraternity (if it had so happened that their power amplifier did not come complete with a pair of speciality speaker cables), even when their total expenditure on their equipment was substantial – and that the position of the channel balance control has been optimally adjusted to provide a central stereo image.

Now, in the interests of tidiness, let us reduce the length of one of the LS cables, so that there is less redundant cable between the power amplifier and the nearer LS unit. Almost certainly it will be found that the position of the channel balance control for a central stereo image must now be altered so that less gain is required from the LS unit fed by the shorter LS lead – a condition which is unaffected by change of amplifiers or LS units. This outcome would be puzzling to the user, if he or she had previously satisfied themselves either by measurement or by calculations based on the maker's specification of resistance per metre, that the resistance of their LS cables was substantially lower than the quoted impedance of their loudspeakers. Experiment suggests that the extent to which the central stereo image position is affected by the relative lengths of the LS cables depends on the goodness (in respect of lowness of resistance) of the cables, so that the offset of central stereo image caused by cable length asymmetry is less noticeable with the better quality cables.

I think that the explanation of this phenomenon is that the LS makers' quoted impedance values tend to be somewhat notional and that the impedance of most multiple-driver loudspeaker units can fall to very low levels at parts of the audio spectrum, especially under dynamic conditions. Therefore the amplifier and its output leads may be required, under some conditions, to drive an LS load of less than an ohm, rather than the 'eight ohms' suggested by the makers.

However, the more enthusiastic members of the hi-fi fraternity, having discovered that cable types can affect the system performance, have made the presumption, on the basis of this discovery, that the fewer signal components carried by the LS cable, the better the system will sound. The logical conclusion from this belief is that the audio performance of the LS unit will be improved if the crossover unit within the speaker assembly is designed so that it can be divided into sections, and so that each section can be individually connected to the amplifier output by way of its individual cable pair. Systems of this kind seem to have gained some popularity, but I feel that such bi-wired or tri-wired set-ups cannot really approach the performance which can be given by active crossover networks feeding individual amplifiers and driver units.

CHAPTER 9

Power supplies

THE IMPORTANCE OF THE POWER SUPPLY UNIT

In the early days of audio, when mains-powered amplifiers were based almost exclusively on the use of thermionic valves, very great attention was paid to the design of the power supply unit from which the amplifier was fed. This was mainly in order to reduce the extent of the 100 or 120 Hz mains 'hum' whose presence, to some degree, was an almost inevitable accompaniment to the audio output of the system.

In more recent times, when low-cost audio systems have been exclusively based on 'solid-state' devices (bipolar junction transistors or ICs) the relatively high degree of power supply line signal rejection – referred to as 'PSRR' (power supply rejection ratio) in the IC manufacturers' data sheets – normally found in solid-state circuitry has encouraged a much more off-hand approach to this aspect of system design. This is a mistake, in design terms, since the output signal, usually developed in respect of an 'earthy' 0-volt line, is applied to the load in series with the power supply unit. This is true, as shown in Fig. 9.1, whether the output is single-ended (as in Fig. 9.1(a)), or 'direct coupled, and balanced about the '0 V' rail (as shown in Fig. 9.1(b)).

Failure to consider the way in which the output current flow path was completed led to a design oversight in a commercial 'pseudo-class A' amplifier which was analysed in an article in *Electronics + Wireless World* (JLH, *EWW*, December 1989, pp. 1167–1168). In the example considered, the power amplifier operated in class A at a high quiescent current, but using a low voltage power supply to minimise the thermal dissipation in the output transistors. The '0 V' line of the low-voltage power supply unit which fed this amplifier was not connected to the main, LS return, '0 V' line, but was taken to the output of a higher power class B amplifier whose function was to move the whole low-voltage supply, bodily, in relation to the '0 V' line. The intention of this arrangement was that the class A amplifier would be able to provide a high LS output voltage swing without a high dissipation. However, this meant that the LS return path to the '0 V' line was now taken via the class B amplifier as well, and any distortion in this would appear in the output to the LS.

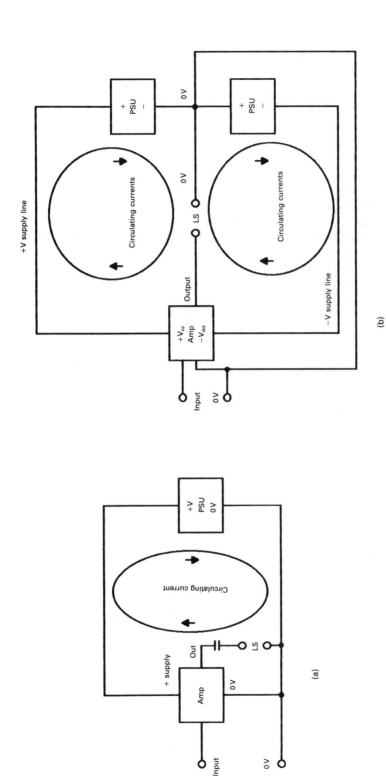

Fig. 9.1 *Amp/PSU current flows. (a) Single-ended system; (b) direct-coupled system.*

CIRCUIT LAYOUTS

The simplest way of converting an input AC voltage waveform of the kind shown in Fig. 9.2(a) into a DC voltage output is to pass it through some device which will only allow current to flow in one direction (normally called a 'rectifier') as shown in Fig. 9.2(c). By the choice of a suitable power transformer and rectifier, any required output current or voltage could be obtained. Unfortunately, as illustrated in Fig. 9.2(b), the output would be a series of uni-directional current pulses. This is satisfactory for operating lamps or motors, perhaps, but not for electronic equipment or audio systems.

This situation is improved if an output 'reservoir' capacitor is included in the circuit, as shown in Fig. 9.3(a), to store the output voltage until the following capacitor charging pulse. However, this is still not a complete answer to the design requirement for a smooth, hum-free output voltage, since the voltage stored in this capacitor will decay as current flows through the external load resistor, as shown in Fig. 9.3(b). The transformer/rectifier circuit can only recharge this capacitor when the peak transformer output voltage somewhat exceeds the voltage on the capacitor, as is the case between the points labelled 1–2, 3–4, 5–6 and so on. This leaves an undesired ripple on the output DC voltage waveform shown in Fig. 9.3(c),

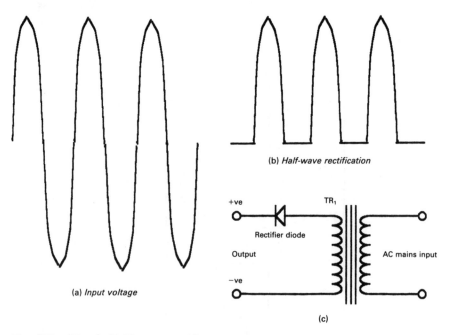

(b) *Half-wave rectification*

(a) *Input voltage*

(c)

Fig. 9.2 *Simple half-wave rectifier system.*

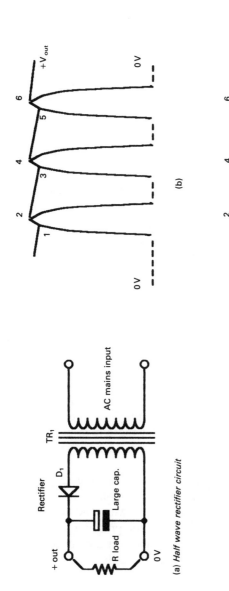

(a) Half wave rectifier circuit

(b)

Output voltage

(c) Half-wave rectification

Fig. 9.3 *Half-wave rectifier output ripple.*

which worsens, for a given size of reservoir capacitor, as the output current demand is increased.

Unfortunately, this recharging current flows only for a brief period each cycle, and it must be larger than the mean output current by the ratio (cycle duration):(conduction duration). If the reservoir capacitor is large, the rectifier diode has a low conducting resistance, and the transformer is of high quality with a low resistance windings then the peak charging currents when the diode is conducting, between points 1–2, 3–4, etc., will be very high.

CIRCUIT PROBLEMS

No electrical circuit has zero resistance (R) at room temperature, and the voltage (V) developed for a given current flow (I) will be $V = IR$. In a typical power supply the reservoir capacitor charging currents could well be hundreds of amperes, and this will cause a significant voltage drop, in addition to that due to the circuit resistances, due to the equivalent series resistance (ESR) in the reservoir capacitor. The voltages produced by these pulsating currents will be added to the existing ripple voltage present on the power supply output line.

A further problem of which the power supply unit (PSU) designer must take note is that the very high short-duration circulating currents, during the recharging of the reservoir capacitor, can generate troublesome voltages across all parts of the wiring (or chassis connections) which are in the path from the transformer to the rectifier, to the reservoir capacitor and back to the transformer again. If any part of the signal chain shares any part of this path, it will have some of this 'spiky' short-duration hum voltage added to it. (A further nuisance is that these high peak currents cause strong, short-duration, magnetic fields, which can induce undesired voltages in adjacent wiring.)

Unfortunately, all the steps that the designer can take to lessen the ripple on the PSU output, of which the most common is the choice of a very large value, low ESR reservoir capacitor, will tend to make the charging duration shorter and thereby increase the pulsating peak current flow. High pulsating currents also make demands on the construction of the mains transformer – an aspect which I will consider later.

FULL-WAVE RECTIFIER SYSTEMS

An immediate improvement in the performance of any simple transformer/ rectifier system can be made if both halves of the AC output waveform are used, rather than only, say, the positive-going half. This technique, in its various forms, is almost universally employed in PSU systems since it halves the duration of the conducting cycle, and thereby also halves the ripple voltage and the peak reservoir (and transformer secondary) pulse currents.

With valve rectifiers, where it is necessary to isolate the cathode/heater circuits, the circuit normally used was of the kind shown in Fig. 9.4(a) for an indirectly heated rectifier valve such as a 5Z4. This kind of valve is to be preferred to a directly heated rectifier, such as the 5U4, if an adequate output current could be obtained, because the HT voltage would only rise gradually following switch-on, as the cathode gradually warmed up to its operating temperature. However, indirectly heated cathodes are less efficient for a given heater wattage, and will have a somewhat shorter operating life.

A fairly common solid-state equivalent system is shown in the circuit of Fig. 9.4(b). The transformer secondary voltage and ripple waveforms for these types of circuit are shown in Figs 9.4(c) and 9.4(d).

With solid-state rectifier systems, in which the designer is free from the constraints imposed by the need for heater connections and supplies, other layouts are possible, such as the full-wave bridge rectifier layout shown in Fig. 9.5(a), used here to provide a single-polarity output voltage. This has an advantage over the layout shown in Fig. 9.4(b) in that instead of each half of the secondary winding conducting alternately, both secondary windings conduct each half-cycle. This reduces the ratio of peak-to-mean current flow in the transformer secondary circuit, and helps to lessen both the audible hum and the electro-magnetic field radiation from the transformer. However, while this reduces the voltage drop in the secondary windings due to the high peak-current flow, it requires the presence of two rectifier diodes in the path from secondary winding to DC output, instead of just one.

For a twin-polarity power supply output, the most usual circuit layout is that shown in Fig. 9.5(b), which shares the advantages of more complete utilisation of the current output from the secondary winding and also that of lower peak current levels. Both these layouts will give the type of ripple voltage shown in Fig. 9.4(d). The half-wave rectifier circuit shown in Fig. 9.5(c) is sometimes found in low-current twin-rail supply systems, where the major considerations are those of simplicity and cheapness. It has the output ripple characteristics shown in Fig. 9.3(c).

The choice of the size of the reservoir capacitor in any of the preceding circuits demands some compromise. The desire to keep the output ripple voltage to the lowest possible level – coupled with the wish, particularly in the case of audio amplifiers, to arrange the PSU so that it can deliver high peak output currents, where required – urges the choice of as large a value of reservoir capacitor as the space and money available will allow. However, increasing the size of the reservoir capacitor causes the current

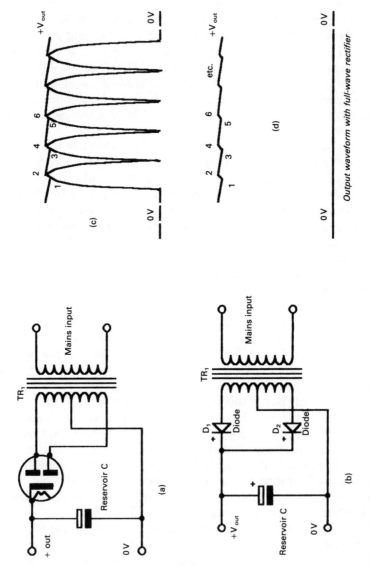

Mains input

TR_1

Reservoir C

+ out

0 V

(a)

Mains input

TR_1

D_1 Diode

D_2 Diode

$+V_{out}$

Reservoir C

0 V

(b)

$+V_{out}$

2 4 6
3 5
1

0 V

0 V

(c)

etc. $+V_{out}$

2 4 6
3 5
1

(d)

0 V

0 V

Output waveform with full-wave rectifier

Fig. 9.4 *Full-wave rectifier systems.*

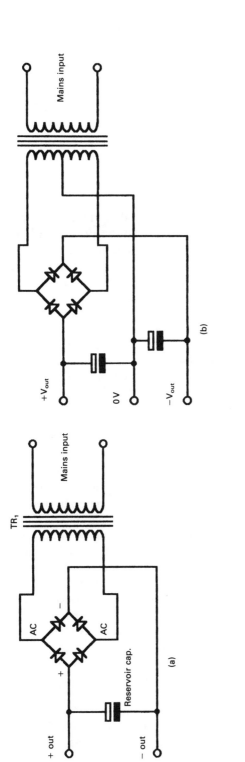

Fig. 9.5 *Additional PSU layouts.*

from the transformer secondary to be drawn in increasingly short pulses, which will worsen the transformer efficiency and heat dissipation. In practice, for any given transformer, the inevitable secondary winding resistance and leakage inductance, associated with core saturation effects, means that there is an optimum reservoir capacitor size, beyond which it is not useful to go.

TRANSFORMER TYPES AND POWER RATINGS

A major pitfall which lies in wait for the unwary PSU designer is that relating to the maximum output current which may be drawn, without overload, from any circuit with an output reservoir capacitor – such as, for example, those illustrated in Figs 9.3(a) or 9.4. The DC output from these layouts will approach that of the peak transformer secondary voltage (approximately 1.4 times the secondary RMS voltage figure, less any voltage drop due to the rectifier or the secondary winding resistance, and less the mean output capacitor voltage droop between recharging cycles).

The transformer secondary winding will then be required to deliver a peak current to the reservoir capacitor which is equal, when averaged, to the mean DC drawn from the system. The transformer is then delivering an output power which is equivalent to $V_{peak} \times I_{out}$ rather than $V_{RMS} \times I_{out}$, and this may be in excess of the manufacturer's rating. A working 'rule of thumb' is that the DC output current (I_{DC}) should not exceed the specified I_{RMS} (max.) $\times$ 0.65.

It is desirable in any mains input transformer that there should be an intimate magnetic coupling between all the windings, and that there should be no stray magnetic field around it. The extent to which this ideal is achieved depends greatly on the nature of the core material (usually grain-oriented silicon steel), and the type of core construction.

In low-cost transformers this core is made by assembling, with the least practicable air gaps in the magnetic circuit, a suitable sized stack of thin, flat, steel alloy 'laminations' – which have been stamped out in the form of 'E's and 'I's – within an appropriately shaped hole in an insulating former holding the windings. The laminations must be tightly clamped to prevent core 'buzz', after which the whole assembly will be dipped in some moisture-proofing and insulating lacquer. In higher-quality contemporary transformers the core will be an annular, 'O'-shaped 'toroid', wound from a continuous thin strip of high-grade magnetic material. The windings are wound around this core by a specially designed machine prior to dipping and impregnation.

Because of the superior magnetic circuit characteristics of the toroidal core, there is much less magnetic field radiation from such a transformer, and it has a superior electrical performance. In high-quality audio equipment these factors will greatly outweigh its less compact construction and its greater cost. Also, because of the way the transformer and its core is made, there is much less core or winding 'buzz' in use – a factor which is welcomed by those with keen ears.

STABILISED PSU CIRCUITS

Low power

There are a number of places in audio amplifier circuitry where a source of some fixed, noise and ripple-free voltage is needed. Some of the circuit layouts used for this purpose are shown in Fig. 9.6. The most common general-purpose circuit is that using a 'zener' diode, as shown in Fig. 9.6(a). (The term 'zener' is used loosely to describe all devices of this type,

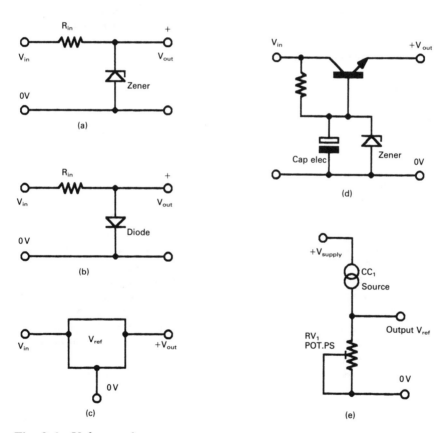

Fig. 9.6 *Voltage reference sources.*

though, strictly, zener diodes are those with output voltages in the range of about 3–5.5 V. Above this voltage they are more properly described as 'avalanche diodes'.) All these reverse-breakdown devices generate a significant amount of electrical noise, and it is usually prudent to connect a capacitor in parallel with the zener to reduce the noise output as, for example, in Fig. 9.6(d).

If the circuit requires only a relatively low voltage output, a much better circuit, from the point of view of noise, is that of Fig. 9.6(b), using one or more silicon junction diodes connected in series in the forward (conduction) mode. Multiples of the forward voltage (usually around 0.57–0.6 V) can be obtained in this way. For high-precision, low-temperature coefficient applications there is a wide range of voltage 'bandgap' reference sources, such as the ICL8069 or the LM369, which offer very high precision, low-noise, reference sources in the range 1.22–10 V, but these would be rather too expensive for general audio use.

A very simple means of providing a small adjustable reference voltage – commonly used to generate the forward bias on the output devices in a transistor amplifier – is that shown in Fig. 9.6(e). The thermal characteristics of the constant current source, CC_1, should be chosen to suit the particular application.

Higher power

Although some manufacturers still prefer to use 'in-house' designs for voltage regulators based on discrete components, the 'three-terminal' voltage regulator ICs of the '78../79..' or '78L../79L..' types have now become an almost inevitable choice for use in relatively low voltage (up to, say, ±24 V) supplies for use with low-level signal circuitry, in preamps or tape/CD/radio systems. The advantages of these ICs are significant. They are cheap, compact, predictable, compensated for variations in ambient temperature, available in both +ve and −ve line versions and have in-built overload protection. They also have a very high degree of power supply ripple rejection, and a low output noise – especially the low-power 'L' series. Although higher-power and higher-voltage versions are becoming increasingly available, they are not, as yet, readily available for use at the voltage, current and power levels needed for audio power amp. supplies. Therefore, in this field, if voltage regulation is required, it will be in the form of a discrete component circuit.

However, the question of whether to use a voltage regulator system opens up philosophical divisions within the audio amp. manufacturers, based on their judgement of their likely market. If the power output stage of an audio amplifier is fed directly from a large reservoir capacitor and a massive transformer/rectifier combination, of the general type shown in

Fig. 9.5(b), it will be able to deliver very high powers into the LS load, especially under short-duration signal demands – the origin of the so-called 'music power' ratings of the 1960s. This capability is particularly welcomed by those for whom the physical impact of high-level sound is almost as important as the artistic content of the signal. For those users any regulator circuit which acted to limit the off-load power supply voltage, and the consequent peak sound level output, would be very unwelcome.

Unfortunately, there are also snags. If the power supply can deliver large output currents at high voltages into the LS load, under short-duration music conditions, it will also be able to do it for a longer time under fault conditions, which could be destructive of expensive LS reproducers. The normal type of damage limitation offered by the manufacturers of amps using unprotected PSUs is to incorporate a suitable 'slow-blow' fuse in the output circuit, coupled possibly with some fast-acting overload 'crowbar' circuit to short-circuit the amp output under overvoltage fault conditions. The snag with the use of any output line fuse is that the integrity of the fuse/fuseholder contacts will deteriorate with time, and will then introduce unwanted resistance into the LS output circuit.

Some rather less obvious design problems are that the residual (100/ 120 Hz) AC ripple on the PSU output will increase as the output current rises, and this will increase the amplifier background hum level at high output powers. This may not be very noticeable because it will be masked by the music output, but it will be measurable under tests using spectrum analysis, and may be commented upon by suitably equipped reviewers.

An additional problem, more apparent to the designers than to the users, is that the off-load output voltage of such a simple supply system can be up to 40% greater than that on load. This means that components in the power amp. will need to support the full off-load voltage, plus some safety margin to allow for mains input voltage surges rather than the supply voltage existing under full-load conditions. These components could be more expensive and, in the particular case of transistors, will usually be of a rather lower electrical performance.

For an audio system which is likely to be used mostly with more traditional music, of which the aesthetic appeal may rely less upon physical impact than upon clarity and tonal purity, a system in which the power amplifier is fed with constant voltage smooth DC supplies, and in which the output power is the same for sustained signal outputs as it is for brief-duration transients, has much to commend it. However, it may be less well liked by the 'heavy metal' or 'reggae' devotees.

Higher power regulator layouts

These can be divided into 'shunt' and 'series' connected circuits, of the kinds shown in Figs 9.6(a) and 9.6(b) respectively, of which the simple

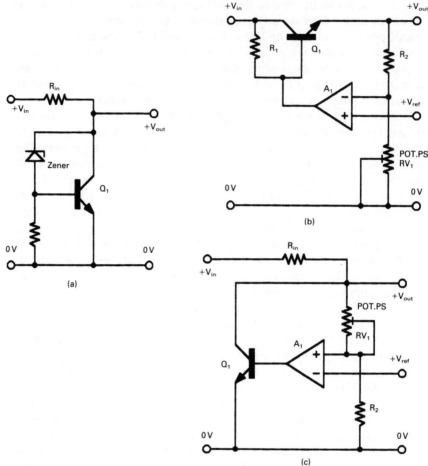

Fig. 9.7   *Shunt and series regulators.*

zener diode circuit of Fig 9.6(a) is the best-known type. Although zener diodes are available with a range of permissible dissipations from 400 mW to some tens of watts, the higher-power versions are not widely available and expensive. The circuit shown in Fig. 9.7(a) uses a combination of a power transistor, Q_1, with a low-power zener diode to give a high-power shunt regulator circuit. The operating voltage of the zener diode needs to be chosen so that the circuit will draw current from the power supply and reduce its output voltage (due to the voltage drop across its source resistance, R_{in}), at some point between its off-load and full power voltage levels.

A more elegant circuit arrangement is shown schematically in Fig. 9.7(c), where the difference between some stable reference voltage and a proportion of the output voltage is amplified and fed to a power transistor connected across the power supply output. The precise operating voltage for this circuit can be chosen by setting the preset potentiometer RV_1.

A typical series voltage regulator circuit is shown in Fig. 9.7(b), in which, once again, a proportion of the PSU output voltage is compared with a reference voltage, and an amplifier, A_1, is used to adjust the current fed into the base of the series 'pass' transistor, Q_1, so that the required output voltage is obtained. The snag with a series voltage regulator of this type is that, for correct operation, the input voltage must exceed the required output voltage by some 3 V – known as the 'dropout' voltage – or otherwise the current flow through R_1 will not be adequate to support the required output current.

In the variations of Fig. 9.7(b) which are available in IC form, the voltage reference is invariably derived from a 'band-gap' type of temperature-compensated voltage source. Not only is the output voltage derived from this arrangement much more stable than that from a zener diode, it is also much more noise free. This is a point which is sometimes overlooked by designers who use a 'quiet' IC regulator to provide a lower voltage to drive a 'noisy' zener diode.

Discrete component voltage regulators can be built to provide any required combination of output voltages and currents, and I am convinced that their use to drive audio amplifier systems does provide worthwhile benefits in sound quality, if for no other reason that such a circuit can have an output impedance which is largely frequency independent and can be less than a tenth of one ohm. To obtain such an output impedance, at, say, 20 Hz from any output capacitor would demand a capacitance value of some 80 000 µF – a very bulky and costly alternative. A further advantage of such an electronically controlled supply system is that it can be made to current limit to protect the circuit components from overload, or cut off the supply completely in the event of any amplifier failure. For my own use, I greatly prefer such systems, and described two such layouts for use with power amplifier designs in *Electronics + Wireless World* (May 1989, pp. 524–527) and *Electronics Today* (May 1989, pp. 25–33), shown respectively in Figs 9.8 and 9.9.

COMMERCIAL POWER AMP. PSUs

At one time, most of the more highly esteemed audio power amplifiers would have employed some form of PSU output voltage regulation. However, this approach has now been largely abandoned, I suspect as a result of market pressures, in favour of simpler layouts whose principal feature is an ability to provide the high peak supply line currents needed for some modern music. The PSU circuit details given in the data sheets for

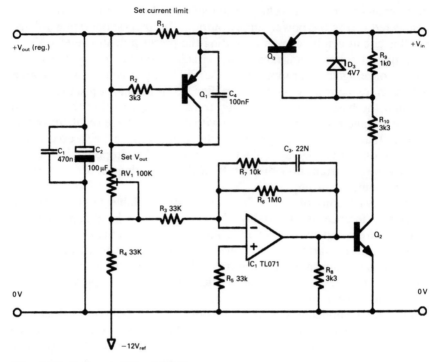

Fig. 9.8   *Series-stabilised PSU.*

the bulk of contemporary power amps show only variations of the simple layout of Fig. 9.5(b), coupled with some overcurrent protection, or relay switching, circuit to give a measure of LS protection.

Nevertheless, some interesting variations still exist, such as the use by 'Pioneer' of a version of the augmented zener diode layout of Fig. 9.6(d) as the main power supply line stabilisation circuit for their M90(BK) flagship amplifier, an arrangement which is also used by Technics and Rotel. An ingenious system shown in Fig. 9.10 is used by 'Quad' (the Acoustical Manufacturing Co.) in many of their current power amp designs for converting a single polarity supply into a pair of symmetrical ± voltage lines.

OUTPUT SOURCE IMPEDANCE AND NOISE

Bearing in mind that the output of any amplifier will drive its load in series with its power supply source resistance, and that a proportion of the noise present on the DC power supply lines may also appear on the output signal, the output impedance and noise characteristics of power supply

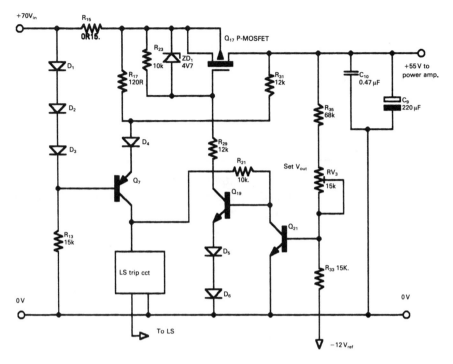

Fig. 9.9 *S/C protected stabilised PSU.*

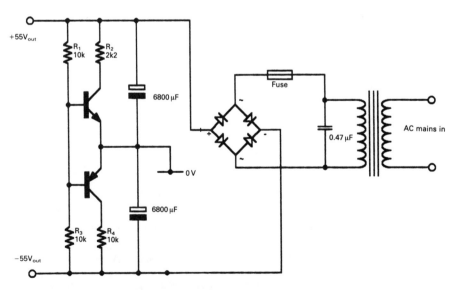

Fig. 9.10 *The Quad power supply system.*

units are of considerable interest to the circuit designer. The commercial three-terminal voltage regulator ICs have a lower noise level and output resistance than small signal zener diodes. They are preferable where these characteristics are important, as, for example, in low-level audio circuitry. Both primary and secondary cells have good electrical characteristics when new, and the performances of these are listed in Table 9.1.

Table 9.1   Cell and voltage regulator characteristics

Cells	Output voltage (new)	Output voltage (used)	Internal resistance (new)	Internal resistance (used)	Noise voltage (μV)
Leclanche	1.55	1.45	1.5 ohms	7 ohms	< 1
Zinc chloride	1.55	1.45	1.2 ohms	3.5 ohms	**
Alkaline manganese	1.57	1.4	0.75 ohms	1.2 ohms	**
Nickel cadmium	1.33	1.25	0.03 ohms	0.05 ohms	**

Regulator ICs	Output voltage (V)	Internal resistance (Ω)	Noise voltage (μV)
7805	5	0.007	40
78L05	5	0.09	40
7812	12	0.008	75
78L12	12	0.18	80
7815	15	0.18	90
7905	−5	0.01	125
79L05	−5	0.15	40
7912	−12	0.01	290
79L12	−12	0.5	95
7915	−15	0.01	375
79L15	−15	0.625	120

| Bench stabilised PSU | 0–35 V | (for comparison) | 330 |
| 400 mW zener diode | 4.7 | | 450 |

Notes. The quoted noise levels were measured over a 20 Hz to 20 kHz bandwidth at 20°C. All the (AA size) battery cells had a noise voltage which was below 1 μV on a 10 mA current drain. The '79' series regulators have a somewhat inferior performance by comparison with the '78xx' types. This is because it is difficult to make good p–n–p transistors in IC form, and this necessitates the use in the '79xx' devices of a somewhat different internal circuit layout.

TRANSFORMER NOISE AND STRAY MAGNETIC FIELDS

All mains transformers generate some audible noise as a result of the alternating magnetic flux and the consequent mutual magnetic repulsion effects within their core and windings, and this can, sometimes, be annoying. 'E' and 'I' type lamination assemblies, used to construct the magnetic cores in conventional low-cost mains transformers, are less good in this respect than the toroidal and 'C-core' constructions used in high-quality equipment. The use of a higher quality core system leads to a much lower external (stray) magnetic field, which facilitates the design of mains-operated units having a low background 'hum' level, and also makes units employing toroidal or similar core transformers less likely to induce mains 'hum' into the circuitry of other units placed in proximity to them.

However, the ability of the core of a transformer to confine the associated magnetic field also depends on the permeability of the core material and this, in turn, depends on the magnitude of the flux within the core. If this approaches saturation the permeability will fall, and the external magnetic field will increase. For this reason the stray magnetic field in proximity to the transformer, as well as its noise level, will increase as the peak currents in its windings increase, and this is a drawback in the use of very large value power supply reservoir capacitors unless this is accompanied by an increase in the power rating of the component.

CHAPTER 10

Noise reduction techniques

Almost all signal handling stages will introduce some unwanted extraneous components. When these unwanted signals are unrelated electrically to the wanted one they are generally termed 'noise', and recording processes are particularly prone to this type of problem. This can arise as a result of mechanical imperfections in the recording process, which can introduce 'rumble' – low-frequency noise characteristic of motor bearings – or 'hum', due to the stray electrical fields from the AC mains supply or from AC ripple, due to inadequate smoothing in the power supply system. Alternatively, and more commonly now that thermionic valves have been replaced by solid state amplifying devices, the noise will take the form of a high-frequency 'hiss', due to thermal (resistor) noise, or to the granular nature of the medium used in the intervening tape recording stages.

BANDWIDTH LIMITATION

The simplest and most widely used technique for noise reduction is just to limit the electrical bandwidth of the system. For example, in the case of the early 78 RPM shellac 'gramophone record' discs the major cause of the hissing noise heard by the listener was the needle tip passing over the particles of emery powder incorporated into the composition of the disc moulding resin to stiffen the shellac mix and to prevent the steel needle cutting its way through the disc. Because this would form the surface of the groove in which the needle was sliding, this powder loading caused a relatively loud 'needle hiss' on replay. However, since the bulk of the energy of this noise was at high frequencies and there was not a lot of useful musical signal recorded above 4–5 kHz, a simple low-pass filter, having a fairly steep cut-off above 5 kHz, would greatly reduce the nuisance value of the hiss without making the replayed music sound too lacking in treble.

The problem of record surface hiss virtually disappeared with the introduction of the (emery powder free) vinyl 'LP' in the early 1950s, but up to that time most of the electronics magazines having an interest in audio topics would periodically offer circuits for new and more versatile needle hiss or 'scratch' filters. (This widely used term was in reality a misnomer, since the removal of the impulse type 'clicks' caused by surface scratches is a much more difficult business, and will be discussed later.)

Unfortunately, the quiet background noise level of a well-made LP, replayed on a well-designed audio amplifier, drew attention to the presence of low-frequency noise components. These LF noise components mainly fell in the 5–30 Hz part of the audio spectrum – sometimes exaggerated by pick-up arm resonances – and could be minimised by the use of a steep-cut high-pass filter with a turnover frequency of 25–30 Hz, with very little loss of wanted LF signals. This is particularly true, in spite of the hopes and beliefs of their owners, because few LS systems or listening rooms will allow the reproduction, with any degree of efficiency, of signals below 35 Hz.

PRE-EMPHASIS

In the particular case of 78 RPM shellac discs, where it was expected that the user would deliberately curtail the HF signal (and hiss) level on replay, some degree of compensation for the resulting treble loss could be provided by deliberately providing pre-recording treble boost, as done in the later RCA 'Dynagroove' recordings, so that the application of subsequent HF (needle hiss) filtration would still leave a more realistic treble signal level. This approach was formally adopted on an international basis by the recording companies, by the use of the RIAA pre-emphasis/de-emphasis format for LP and EP reproduction see Fig. 3.3 and pp. 123–124.

Any substantial degree of pre-emphasis will bring with it the danger of overloading the recording medium, but, in the case of the gramophone record, this risk is minimised by choosing the frequency of the treble pre-emphasis so that there is very little treble lift below some 2 kHz, since above this frequency it is not expected that programme signal components will be very large.

'NOISE MASKING' AND 'COMPANDING'

Where the curtailment of the reproduced audio bandwidth is not an acceptable solution to the need for noise reduction, other more sophisticated techniques can be employed, such as the reduction of the size and dynamic range of the signal prior to the recording process, followed by a compensatory increase on replay – a process called 'companding' (**comp**ression plus exp**anding**).

Unfortunately, the proper operation of this process will lead the user into a technical minefield, in which a number of contributory problems and effects must be recognised. Of these, an important phenomenon is that of 'noise masking'. This is a basic feature of the human hearing process, and relates to the acoustic effect that a loud signal can prevent the listener from hearing a quieter signal present at the same time. This masking effect is particularly noticeable when the louder signal is at a frequency close to that of the weaker one, when the hearing threshold is raised (i.e. the loudness of the signal must be greater before it is heard) in the neighbourhood of the frequency of the louder one.

One effect of this, in relation to wide-band, low-level noise, is that if there is a loud signal present at the same time, the listener will not be aware of any low-level background 'hiss'. This is particularly true if the main signal is distributed over a wide frequency spectrum – as, for example, in the case of a pop group or a symphony orchestra – rather than if the main signal is, say, a single instrument whose output frequency is some way removed from that of the noise background.

This leads to the possibility that if the wanted signal can be increased in loudness before it is recorded, and then reduced in loudness on replay (a process called linear companding), the relative audibility of any noise introduced during the recording process will be reduced, in comparison with the wanted signal. This could be done, quite simply, if the recording engineer turns down his recording gain control when any loud passage occurred, and then turns it up again during the quiet bits. One would then need to arrange the listener to turn up the volume control of his hi-fi equipment on all loud passages and then turn it down again during the relatively quiet ones to reduce the size of the background hiss.

ATTACK AND DECAY TIMES

This simple manual companding process would obviously be impractical in real life, but it could be done, automatically, by the use of suitable electronic circuitry, as long as a mutually agreed signal level had been chosen as the turn-over point for the onset of compression and subsequent expansion. (Normally, this problem is evaded by simply allowing the companding to continue down to zero signal level, but this is not very efficient.) Unfortunately, the problem would remain that in order for the system to perform efficiently the loudest signal fed from the compander to the recording system would have to be close to – but not beyond – the maximum permitted level. This then leads immediately to the problem of the 'attack' time of the compression circuit. Let us suppose that we wished to improve the signal-to-noise ratio of the recording process by $4\times$ (12 dB), then we would need to increase the size of a low-level input signal to the recording system by 12 dB while ensuring that high-level signals did not ever exceed the ('0 dB') signal overload limit.

There are several difficulties here, of which the first one is that no gain adjustment system can ever work instantaneously. The second one is that, even if it could, it is essential that all of the signal modification processes which are used must be inaudible to the listener. A rapid compander action would fail this requirement since, if the speed of response – the attack time – of the gain control circuitry is fast, its action in suddenly altering the signal level to the recording system would be audible as a series of 'clicks', while if the attack time is slow, the recording system will be driven into overload during the time which it takes the gain control system to reduce the gain by 12 dB. The corollary to this is that, if the incoming signal should suddenly be reduced in size, there will be a time lag before the recording gain level is increased once more to the required low-level signal (+12 dB) target value.

SIGNAL LEVEL LIMITING

An inevitable back-up stage in any recording signal conditioning system is a fast-acting limiter stage. This need is imposed by the fact that any recording (or broadcasting) process has finite upper amplitude limits which must never be exceeded, and the incoming signal level must be prevented from over-running this limit. Amplitude limiting devices are of two varieties – 'soft' limiters in which the signal peaks are progressively rounded off as they approach the limit level, and 'hard' limiters or clippers, which are, as their name suggests, circuits which electrically square off the peak amplitude of the signal waveform so that it is never greater than some predetermined level.

Although, understandably, the soft limiters are said to be preferable, sonically, recording engineers accept that even hard clipping may be inaudible to the listener provided that the clip duration is kept short; less than 1 ms seems to be an acceptable figure. The proportion of the waveform peaks which is clipped is evident by a progressive hardening of the quality of the sound. Insofar as the use of a soft limiter will allow the signal to be recorded at a rather higher level without significant degradation of the sound quality, this will also permit an improvement in the unmodified signal-to-noise ratio.

GRAMOPHONE RECORD 'CLICK' SUPPRESSION

An endemic problem with the old shellac discs was that being brittle, they could crack and then, on replay, there would be an irritating click in synchronism with the rotation of the disc – and this problem remains due to surface scratches even on vinyl EP or LP discs, because of the ease with which surface blemishes can arise after only a few playings. These blemishes are due either to small surface scratches, small depressions due to dust having been impacted onto the disc surface, or to the collapse of the vinyl skin which had covered a small gas bubble formed during the moulding process. Whatever the cause, the remaining 'ticks' and 'pops' can irritate the listener, and circuits have been proposed, from time to time, to alleviate this problem (see Linsley Hood, J.L., *Wireless World*, January 1983, p. 48, Van der Wal, D.J., *Electronic Components and Applications*, August 1980, pp. 215–218).

Neither hard nor soft limiting, on its own, will do more than merely lessen this nuisance, while more elegant methods demand some form of impulse waveform recognition and anticipation. In these latter circuits, the technique employed is to use a fast response voltage spike detector to trigger a high-speed electronic switch so that it either clips or mutes the audio channel, ideally for the duration of the noise spike. In the case of the circuit proposed by Van der Wal, a four element Sallen and Key filter is connected in the audio channel to generate a 5 μs delay, to allow the muting circuit to anticipate the arrival of the noise impulse. A steep cut 20 kHz LC filter of the type shown in Fig. 6.5 would generate a 60 μs delay, which might be more useful. However, experience has shown that none of these click suppression circuits will do more than reduce the conspicuousness of the defect.

PROPRIETARY NOISE REDUCTION SYSTEMS

In the early 1960s there was a considerable interest in techniques which would allow a reduction in the background hiss which spoiled the performance of the otherwise excellent professional reel-to-reel tape recorders, and a number of add-on or build-in proprietary systems were offered. Of these the more noteworthy were the 'Telcom C4' process evolved by Telefunken, the 'dbx' system, and the various 'Dolby' variants evolved by the Dolby Laboratories – a firm specialising in tape-noise reduction circuitry. Of these alternatives, from the layman's point of view, the various Dolby systems (so far these are the Dolby 'A', 'B', 'C', 'S' and 'SR'), have been the most successful, and the most widely adopted. An excellent review of the various noise reduction systems is given by Fisher (Fisher, D.M., *Newnes Audio and Hi-Fi Handbook*, ISBN 0 7506 0932 X, Chapter 8).

Dolby A

This was first introduced, commercially, in 1966 as a method of improving the signal-to-noise ratio of wide-bandwidth, professional reel-to-reel recorders, and employed – in a similar manner to the contemporary Telcom C4 process – a 'complementary' companding system which operated independently on four different frequency bands in the incoming audio spectrum. (80 Hz LP, 80 Hz–300 kHz BP, 3 kHz HP and 9 kHz HP). This is an arrangement which allowed better performance than a compander, which operated on the whole bandwidth of the incoming signal, for reasons which will be explained. (I have used the term 'complementary' to denote an arrangement in which the degree and gain/frequency profile of the post recording signal level expansion is, as nearly as possible, a mirror inverse of the treatment given to the incoming signal prior to the recording process.)

The innovative feature introduced by Dolby was the 'Bi-Linear' companding system whose way of operation is illustrated in Fig. 10.1. This avoided the problem of overload in succeeding stages (e.g. the tape recorder) due to the inevitable time delay in the operation of the gain-reducing circuitry following the appearance of a high level signal.

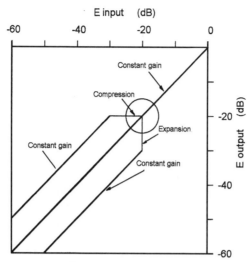

Fig. 10.1 *Bilinear companding*

As shown in Fig. 10.1, for signal levels below the 'rotation point' (ringed) the incoming signals are compressed, which allows a higher effective recording level. Similarly, on replay, signals below this level will be expanded – by 10 dB in the example shown. The basic problem that a large incoming signal could cause the stages following a compressor to overload – during the time it took the internal gain control system to respond to the 'reduce gain' command – is thereby avoided by not allowing the compressor stage to operate at signal levels above the rotation point value.

The value of dividing the frequency spectrum of the incoming signal into discrete segments before processing is that it avoids the problem of 'noise modulation', a term given to the audible effect in which the level of any background 'hiss' becomes louder or quieter in synchronism with the loudness of some signal at a very different frequency, where noise masking did not occur. This ebb and flow of the noise background, as the signal varies in size, is sometimes called 'breathing'. Band division also helps prevent the gain of the compander being affected by 'out of band' signals which will not be recorded anyway.

There is, however, an inherent problem with the use of bi-linear companding which does not occur in a normal wide-band linear compander, and that is that it is essential to ensure that the gain setting used at the 'rotation frequency' is the same in both the compression and the expansion processes. If these are not matched, the frequency and gain balance of the reproduced signal, after processing, will be in error. The solution proposed by Dolby was to specify a 'Dolby Level' for the amplitude and frequency of a 'standard' signal. For easy audible identification this used a 0VU 'warble tone', frequency modulated between 850 Hz and 930 Hz, and this would be recorded at the beginning of any Dolby encoded tape, to allow the replay level to be set. The target wide-band noise reduction for the Dolby A system is 10 dB.

Dolby B

Although the 'Dolby A' noise reduction process proved very successful in use, and was widely adopted by the professional recording fraternity, it was thought to be too complex and expensive (bearing in mind the need for a separate bi-linear compander and filter for each segment of the incoming signal spectrum) for widespread use in low cost domestic tape recorders – a category which was coming increasingly to mean the stereo cassette recorder. This was demonstrably in need of some noise reduction process to improve the wide band s/n level from the typical 50 dB of a cassette replay (ref. the typical '0 dB' figure of the recording level meter) to compete with the 60–65 dB replay s/n ratio of a good vinyl LP measured over the same dynamic range.

The technique proposed by Dolby was to employ a sliding band compression/expansion arrangement in which the extent of the compression of an incoming signal will depend both on its size and on its frequency, as shown in Fig. 10.2 for the decoder. The same process is applied, in reverse, in the signal encoding stage for a Dolby B processed pre-recorded tape. The arrangement by which this is done is shown in Fig. 10.3, and the system gives a reduction in the background wide-band noise level of 10 dB. i.e. a replay s/n ratio of perhaps 60 dB.

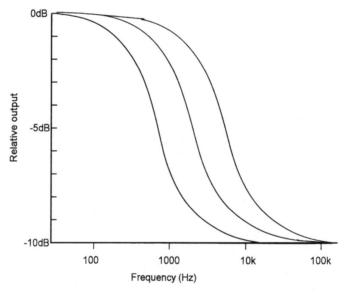

Fig. 10.2 *Dolby B decoder response*

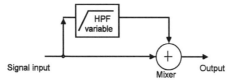

Fig. 10.3 *Encoding unit*

Since it was envisaged (correctly) that Dolby processed pre-recorded cassette tapes would become the normal commercial standard, it was necessary – as in any bi-linear encoding/decoding arrangement – for the rotation point (the frequency of change-over from one pre-emphasis slope to another) to be specified, and this was done by stipulating this level, called the 'Dolby Level', in terms of the remanent flux level on the tape. The value chosen was a fluxivity of 200 nano-Webers/metre. The cassette player will generally have the correct replay level as a pre-set value, though some up-market hi-fi units will provide a built in two-tone generator to enable the gain and bias settings to be correctly chosen by the user, so that his own recordings will conform to the Dolby standard.

The status of the Dolby B noise reduction system as the pre-eminent, and virtually the only, system used in domestic cassette recorder units was underlined by the integrated circuit manufacturers, such as National Semiconductor Corporation, who provided application-specific devices, such as the LM1131, as bilinear stereo encode/decode components matched to the specifications of the Dolby B system.

Dolby C

This was introduced – specifically as a cassette tape recorder improvement facility – in 1980, and is, in simple terms, two Dolby B processor stages connected in series, each contributing a low-level s/n ratio improvement of 10 dB, but operating at different rotation frequencies. This means that their effect is spread out over a further two octaves, and a gives a further 10 dB in noise reduction, as shown in Fig. 10.4. The final result, shown as the extent of noise reduction as a function of frequency, for the Dolby B and Dolby C processes, is shown in Fig. 10.5.

DIGITAL SIGNAL PROCESSING AND NOISE REDUCTION

It is typical of progress in engineering that parallel developments in allied fields tend, in time, to undermine or supplant existing systems, no matter how carefully and laboriously these have evolved to perform some similar functions in an

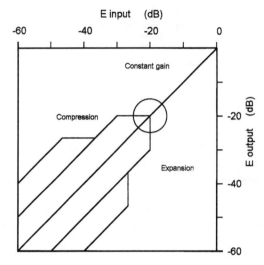

Fig. 10.4   *Effect of Dolby C process*

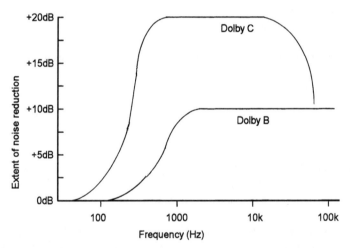

Fig. 10.5   *Dolby B/C decoder response*

alternative domain. An obvious example of this technical jostling for the lead in the field of audio is provided by the analogue vs. digital contests of the present date (1998).

The vinyl LP and EP 33/45 RPM discs, mastered on analogue tape recorders having an ultimate wide-band s/n ratio of perhaps 80 dB, but degraded to some 65 dB on disc replay, have now been almost entirely supplanted by compact discs, whose sleeves display the symbol 'DDD' to show that they have been digitally recorded and mastered and presented in this form to the user, and whose s/n ratio, over a similar dynamic range, is limited to some 80–90 dB by the studio signal handling electronics and the background acoustics of the recording stage. Multitrack analogue tape recorders, with highly refined noise reduction processing, are still used, but they are a dying species in a field where the signal may be handled entirely in the form of solid state memory.

A similar situation now exists in respect to the cassette recorder. The Philips 'Compact Cassette' is capable of a replay s/n ratio of perhaps 70 dB in a carefully set-up Dolby 'C', 'S' or 'SR' equipped cassette player (the latter two Dolby systems are developments of the 'C' process to lessen the problems, such as HF saturation, due to the tape itself) but the average user sticks to his Dolby B machine, because both it, and the tapes it uses, are cheap to buy, even though it has an s/n ratio, at the best, of some 60 dB and a bandwidth of 30–15000 Hz.

For the true hi-fi buff, wishing to make his own recordings, the choice will lie between a re-recordable 5.25 inch CD or a similar recordable 'mini disc', with the odds currently stacked in favour of the (all digital) recordable CD, mainly because this can be replayed on a standard CD player, and for which no analogue noise reduction circuitry is needed. The self-appointed audio connoisseurs may lament the passing of the LP, with its 'warmth' of tone, but the contest is like that of the sailing ship vs. the jumbo jet. The schooner may be more fun to travel on, provided the weather is calm, but if you want to get from London to Seattle in a hurry, you will travel by air.

CHAPTER 11
Digital audio broadcasting

Broadcasting system choices

AM

Until the 1990s, there were only two basic options for broadcasting a radio signal – systems in which the signal is modulated in amplitude (AM), and those using frequency modulation (FM) – and each of these methods have their individual advantages and drawbacks (see Chapter 2). In the case of AM, which is the longest established, and still the most widely used method, its advantages are its cheapness and simplicity of construction, in respect of both the transmitter and the receiver. It also allows by far the most efficient utilisation of the available electromagnetic spectrum. For example, it is possible, in principle, to broadcast a (mono) AM signal carrying modulation frequencies covering the range 20 Hz to 20 kHz with a transmission bandwidth of $f_0 \pm 20$ kHz, where f_0 is the 'carrier' frequency.

Unfortunately, AM broadcasting also suffers from a large number of problems in use. To begin with, there is an effect which will be obvious to any user, in which the strength of the signal received by the same receiver from the same transmitter will vary from season to season, from daylight to darkness, and even from minute to minute (a phenomenon known as 'fading') because of the variable degree of refraction of the radio signals by the ionised upper regions of the atmosphere through which they pass, as shown earlier in Figs 2.2–2.4. Also, more distant and less powerful transmitters will offer a weaker signal to the listener, so that the tuning dial of his AM radio will be crowded with a mix of signals – some weak, some strong – and if a strong signal happens to be close, in its carrier frequency, to the one selected then there will be a spillage of programme from the strong into the weak, a common problem known as 'adjacent channel interference'. Increasing the sharpness of tuning of the receiver will reduce the tendency of a powerful adjacent channel signal to break through, but only at the cost of progressively restricting the ability of the receiver to reproduce the higher modulation frequencies of the transmitted programme.

VHF FM

In this technique, the size of the carrier is held constant and the required modulation is impressed on it by varying the instantaneous value of the carrier frequency. In the receiver, the chosen signal is then amplified and amplitude-limited so that all the received signals will have the same final size. Since all the signals will be as big as each other, such an arrangement will be, at least in theory, highly resistant to the intrusion of amplitude-modulated 'impulse noise' such as that generated by motor car ignition systems. This allows FM broadcasts to be made in the VHF frequency bands, where AM broadcasts are plagued by electrical interference from passing motor traffic.

Other advantages follow from the choice of a VHF broadcasting frequency, of which the major ones are that transmissions at such frequencies are much less affected by the atmospheric ionisation layers, and will therefore be restricted to line-of-sight propagation paths. This will, in turn, greatly reduce the number of possible interfering transmitters. Also, since the VHF bands are much less congested with radio traffic, their use allows much more elbow room between transmissions, and this permits the broadcasting of a low distortion L/R stereo pair of signals having a good freedom from interference and background noise, and a modulation bandwidth of 30–14,000 Hz, although to do so requires some 250 kHz of RF bandwidth.

This assumes, however, that the receiver is sited in a fixed location, and has a good aerial – which often means that it will be fairly bulky. If the receiver has a small built-in aerial, as would be the case for a portable 'trannie', the physical position of the radio set, and its location in relation to other objects, can greatly affect the strength of a wanted FM signal. Even someone walking about in the room near the radio can cause the signal to fade or distort, or even on occasions to disappear entirely. This problem is particularly annoying to the user of an FM car radio where the reception of the wanted signal can be spoilt by a wide range of reflecting objects, and makes the use of an FM car radio, on a long journey by road, an irritating exercise requiring constantly re-tuning from one fading and distorting signal to another.

Digital audio broadcasting (DAB)

THE EUREKA-147 PROJECT

A proposition of my own is that the influence which a person can exercise is directly proportional to the extent to which they travel. The implication of this is that the users of car radios will be an influential section of the community. The recognition of the existence of such a numerous, probably affluent, and politically important section of the population led, in 1985, to the creation of a forty-two member consortium including broadcasting authorities, university research departments, motor car and car radio manufacturers and other interested parties, under the code name 'Eureka', ('Eureka-147' refers to the serial number of this particular DAB project), to address the problems of car radio reception – with particular reference to the possibilities for better reception offered by the use of digitally encoded radio signals.

Undoubtedly, the challenge to be met was that of the audio quality obtainable from the compact disc, and from the CD players which were increasingly being fitted to luxury cars. However, although it would be possible to broadcast and receive the RF signal generated by a CD player, and indeed this was done, experimentally, in the USA, in 1986, it would be extravagant, at some 4.3 MHz per channel (see Fig. 6.6), in its use of radio broadcasting bandwidth, and it was felt that this could be improved upon.

In particular, the use of digital encoding will permit advantage to be taken of the (relatively) very high operating speed of digital electronics, so that, for example, at the proposed sampling rate of 48 kHz, there is an interval of 20.8 µs between the time at which each successive sample is taken, and many digital operations can be carried out in such a long interval – such as the correction of received errors and the evaluation and manipulation of the characteristics of the sample which has just been taken.

These possible operations are combined in the 'E-147' system by a sampling process called Coded Orthogonal Frequency Division Multiplexing (COFDM). In this, the term 'coded' refers to the fact that, as in a CD replay, the received signal has been transformed, before broadcasting, by convolution and inter-leaving and by EFM (eight to fourteen bit-length modulation) to clean up and allow error correction in the final pulse train, while the FDM part of its title means that the transmitted signal is broadcast and subsequently sampled, in the receiver, as a multiplex sequence of up to 1000 individual 1.5 kHz wide carrier frequencies, spread across a 1.5 MHz frequency band, each of which is sampled in a predetermined order.

DIGITAL RADIO AND TV: THE GROWTH OF COMPLEXITY

All the earlier chapters of this book, with the exception of Chapter 6, have been concerned with the traditional analogue-based electronics used in audio applications, because this still represents by far the largest part of consumer hardware, and is still accessible to the amateur user. However, during the past twenty years, there has been a growing interest in the use of digital audio technology – thrust into public awareness by the appearance in 1982, almost in its final form, of the CD player. Since then there has been an enormous growth and development of digital audio technology (though mainly not within the public gaze), so that there is now as big a gulf in techniques between a conventional AM/FM radio and a digital receiver as there is between the electronic complexity of a vinyl disc replay system and that of a CD player.

A lot of this DAB technology is similar in its nature and action to that explored in Chapter 6, but there remains an important difference between the DAB and the CD, in that while the designer of a CD player can reasonably assume that the data on the disc will be free from major corruption, as played, in the case of a digital audio (or TV) broadcast there is a strong presumption of reception interference, and there is a great interest, in consequence, of robust transmission/encoding and reception/decoding techniques.

PHASE SHIFT MODULATION SYSTEMS

Every digitally encoded signal exists, in its own time frame, in relation to some frequency-stable clock pulse – either received as part of an incoming signal, or generated locally, in synchronism with the incoming data stream. This allows a modulation/demodulation process in which, for example, the '0's and '1's of the data stream will be represented by the phase of the received pulse at the moment of transition of the clock signal, so that a binary '0' would be the receiver response to a received signal pulse which was in phase with the clock, while a binary '1' would be the response to an incoming signal which was in antiphase: a process named 'binary phase shift keying' (BPSK). However, this can be elaborated further to make use of the 90° and 270° pulse/carrier phase relationships – known as 'quadrature phase shift keying' (QPSK).

Obviously, this type of phase shift modulation could be elaborated still further to make use of the 45°, 135°, 225° phase sequences, etc., but the choice of the QPSK system was made because this combines a higher data transmission rate with the lowest bit error rate (known as the BER). An error rate of 10^{-4} is sought.

To avoid confusion at the moment of the shift in carrier phase, a 'guard interval' is interposed in the receiver on either side of the transmitted symbol – bearing in mind that, at the receiver, each incoming symbol will be received in the form of a short burst of RF sinewaves. This period during which the receiver is electronically muted is also of value during conditions of RF multiple path or echo reception, as will be seen later.

There is an inherent advantage in digital signal transmission following from the facility for amplifying and squaring-off the incoming pulses, in that this gives about a six times improvement in the receivability (carrier to noise ratio) of weak DAB signals as compared to that in FM reception.

MULTIPLEX (MUX) SYSTEMS

One of the very attractive features of the compact disc is the facility which it offers for the correction of errors, the kind of replay faults which would have been responsible for the 'clicks' and 'pops' caused by dust and minor scratches on the surface of a vinyl disc and which intrude upon the desirably silent background of an LP or an EP. In the case of the CD the error correction is achieved by deliberately convoluting the signal prior to recording so that when, following replay, the correct order of samples is restored the longer duration 'burst' errors will have been broken down into individual bit errors, which can be corrected by incorporating a measure of redundancy in each recorded word –

a process termed Cross-Interleave Reed–Solomon Coding (CIRC). A version of this error correction process is also used in the E-147 scheme.

In the case of the E-147 system, the transmission consists of a group of up to 1000 digitally encoded audio or data signals, interleaved in frequency and time across some segment of the 7 MHz allocated carrier bandwidth, whose error correction system is based, in part, on the assumption that the error is unlikely to have affected all of the possible pseudo-randomly selected samples at the same time. As in the case of the Sony 'Mini Disk', 'perceptual coding' (known under the acronym 'MUSICAM') is employed to reduce the amount of data used to record the signal.

PERCEPTUAL CODING

In its essentials, perceptual coding operates by analysing the whole audio signal and deleting all those parts of it which are deemed to prove inaudible because of their quietness, or closeness in time or pitch to some other louder signal component present at the same time.

(Personal note: I have some misgivings about this reliance on presumed inaudibility to allow data reduction by the removal of part of the audio signal. My reservations arise from the experimental finding that relatively small, and inaudible, signals will suddenly become audible, by causing a change in timbre of a dominant pure tone, if the previously inaudible small amplitude tone should be in frequency coincidence with some harmonic of the larger one. This implies that if these small tones are deleted because they are deemed not to be audible, the major tones may be shorn of part of their tonality.)

A further facility which is implicit in the multiplex sampling process is that not all of the final 'channels' (each of which would appear to the listener to be a different and distinct 'radio station') put together from the multiplex output, need have the same digital resolution, so that a symphony concert might, perhaps, be encoded as the maximum available 384 kb/s resolution, while a spoken commentary for football might only be allocated the minimum 48 kb/s of digital space.

The final performance offered by the E-147 process, according to an evaluation by Stirling (Stirling, A.M., *IEE Review*, Jan. 1998, p. 14), is a dynamic range of 40 dB, a relatively high level of quantisation noise and an AF bandwidth of 15 kHz. If this criticism is valid, the system would obviously meet the requirements of an improved car radio receiver but not those of a high quality domestic hi-fi installation.

AVOIDANCE OF TIME DELAY DISTORTION

A major problem with any VHF transmission is that reflections can occur from relatively small objects, such as, perhaps, another vehicle in proximity to a car radio user, within line of sight of both the transmitter and the receiver. Even more troublesome, because the signals are larger and the difference in path lengths are normally greater, are those reflections which arise from ranges of hills or other large physical features. Such reflected signals will be delayed in time of arrival in relation to that of the wanted signal, and this causes audible distortion in the received signal because of the waveform modification which results when these direct-path and delayed signals are added together.

In the reception of a digitally encoded signal, this problem can be avoided by making the sampling 'window' large enough so that all of the signals received within a realistic interval following the moment of sampling are added together before decoding.

DAB TRANSMITTER AND RECEIVER LAYOUTS

The schematic layouts of a DAB transmitter and a similar receiver, according to the E-147 format, are shown in Figs 11.1 and 11.2. In the transmitter layout shown in Fig. 11.1, the main multiplex division is controlled by two separate inputs, of which the first is termed the FIC (fast information channel – so-called because the data within it are not subject to the delays due to processing procedures in the programme channels), and which is the path for the service and multiplex structure commands. The second of these is the pathway for the audio and associated information data which are gathered together by the MSC (main service channel) multiplexing circuitry. The combined multiplex sequence from the COFDM is then caused to phase modulate a constant amplitude VHF carrier.

By comparison with this, the basic structure of the E-147 receiver, shown schematically in Fig. 11.2, is relatively straightforward. In this, the broad-band

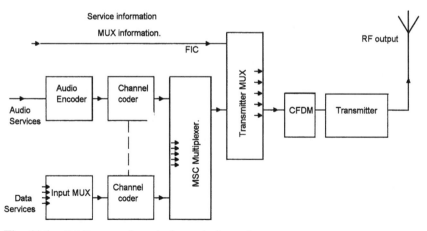

Fig. 11.1 *DAB transmitter (schematic layout)*

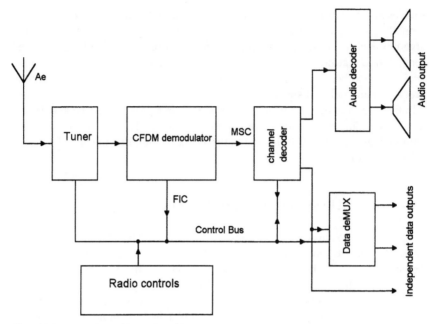

Fig. 11.2 *Eureka-147 DAB radio*

VHF signal, in digital form, is amplified and passed to the COFDM receiver multiplex, from which the user makes the choice of signal which is received – whether this be audio or data – by means of the channel decoder system, and the selected output is converted back into an audio signal, where appropriate, by a conventional error correction and D/A decoding followed by a stereo audio amplifier of much the same form as that shown in the right-hand half of Fig. 6.9. The similarity of the decoding process needed for the E-147 system to that used in a CD player means that many of the same ICs can be used for both arrangements.

Other features of the E-147 system are the ability to change from channel to channel – the 'station', in the sense of the transmitter, will be the same for the whole service area – by the choice of the output, or group of outputs, from the main receiver multiplex, which is part of the COFDM demodulator. Since the national broadcast network will use the same multiplex grouping, this means that one can travel through the whole area of the national network without need to 'retune' the receiver. Also, the data channels shown in Fig. 11.2 allow an accompanying video system to display both text and pictures – although the safety objections raised in the context of in-car TV are likely to restrict the data display to programme and channel information.

THE NICAM-728 TV STEREO SOUND SYSTEM

The 'PAL' colour TV system, used almost exclusively in Europe, and also widely adopted by other non-European national TV systems, incorporates a very high quality mono sound system, of which the 6 MHz 'carrier' is a constant amplitude difference frequency signal, which exists as the heterodyne between the sound and picture carriers – as shown in Fig. 11.3. This is used to provide an a audio output by the use of an FM radio system of conventional structure, shown in the block diagram of Fig. 11.4. The 6 MHz FM input signal for this is extracted from the output of the (negatively modulated) AM picture demodulator stage.

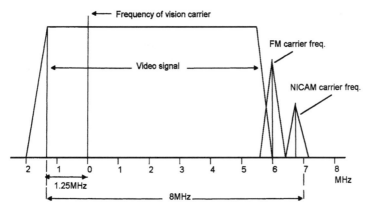

Fig. 11.3   *Spectrum of TV signal*

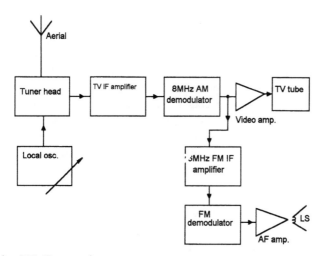

Fig. 11.4   *PAL Tv sound system*

Unfortunately, at the time the PAL system was evolved, it was not thought necessary to make the system provide a stereo sound output, and ever since the introduction of the GE-Zenith FM/TV stereo sound system (see pp. 66–70) studio engineers have regretted the absence of a high quality stereo sound to accompany the picture. The 'NICAM 728' system is a projected answer to this lack.

(Note: there is a psycho-optical problem in the use of stereo sound to accompany the picture. When in 'mono', the brain is quite happy to accept a single source location for the perceived action – for example, with the sound of ball on racquet in a tennis match, the sound may appear to come from either side of the wide TV screen, while the brain knows quite well that the sound is actually coming from a small loudspeaker located to one side of the tube. If the sound is in stereo, the studio engineers must try to ensure that what the eye sees is what the ear hears, and this may present problems.)

The penalty inherent in any change to an existing public service system, such as a TV broadcast network, is that it must be 'backwards compatible'; in other words, it must allow the users of existing TV sets to continue to use the existing broadcasts, without apparent modification to, or degradation in, the quality of their TV sound or picture. In the case of NICAM-728 (the acronym is derived from 'Near Instantaneous Companding Audio Multiplex', and the '728' refers to the fact that the information transfer rate of the digital bit-stream is 728 kbits/s), advantage is taken of the very high reception sensitivity of a radio transmission broadcast in digital form to allow a small part of the 8 MHz wide TV channel to be stolen for the subsidiary sound channel, centred on a carrier frequency 6.552 MHz above the main TV picture carrier.

Unfortunately, as noted above in the case of the compact disc, converting an audio signal from analogue to digital form is very extravagant of bandwidth, and this injected NICAM signal must not interfere with either the sound or the picture on the channel to which the TV is tuned, or with an adjacent one. As shown in Fig. 11.3, there is some residual channel overlap of the TV picture modulation and that of the TV sound and Nicam additions, and this would cause some patterning of both the chosen and adjacent TV picture channels and a worsening of the background noise level of the sound. This is alleviated by 'scrambling' the sequence of the digitally encoded NICAM signal, on transmission, and unscrambling it again, on reception, in a pattern controlled by a pseudo-random number generator. The worsening of background noise is alleviated by fixing the amplitude of the NICAM carrier at -20 dB (one tenth) with respect to the main picture signal, and partly by noise shaping (see pp. 258–9).

THE NICAM-728 AUDIO SIGNAL

This is a straightforward type, which has many similarities with the digital signal in a CD system, and, as derived from the phase modulated NICAM carrier by a

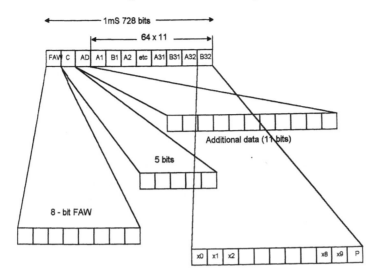

A1 - B32 = 64 11bit sound and parity words

Fig. 11.5　*NICAM 728 bit sound frame*

normal QPSK demodulator, consists of a digital bit-stream organised as shown in Fig. 11.5.

Experience over the years has shown that '14 bit' encoding is about the minimum resolution tolerable for truly hi-fi sound quality, though the BBC only uses 13 bit encoding for its inter-transmitter sound distribution system. This, with some inadvertent 'noise masking' of the quantisation error by the inevitable background hiss of the electronic signal handling stages, has apparently been generally accepted as adequate by the hi-fi fraternity. However, as already noted, the bandwidth requirements of a digitally encoded audio signal increase rapidly with each additional bit of resolution, and this leads to a problem, as explained below.

The NICAM data reduction process begins with an audio signal which has been encoded in a 14 bit (16384 step) digital form. Since the sample rate is 32 kHz, this allows an audio passband of 30–14,500 Hz, and a s/n ratio of about 80 dB. Unfortunately, the maximum resolution which can be accommodated within the available NICAM bandwidth is 10 bits (1024 steps), so some compression is essential. This is achieved by breaking down the input data stream into 1 ms blocks, each containing 32 samples. These are then recoded to 10 bit resolution using a scale factor determined by the largest signal within that block. The scale factor used, within one of five gain ranges, is then indicated by a 3-bit data block added to the control signal, so that the appropriate output scale factor can be restored when the 10 bit signal is decoded.

The format of the 728-bit sound frame, shown in Fig. 11.5, is arranged so that there is an initial 8-bit frame alignment word (FAW), followed by a 5-bit control information word and an 11-bit data block for use where further system control requirements arise. A final group of 704 bits then carries the sound and parity bits arranged as 64 × 11 bit groups in which the L and R channels are alternated.

In order to prevent transmission errors – due, perhaps, to a prolonged noise pulse – from being carried over into successive data blocks, each 64-bit block is interleaved with its neighbours to give a 16 frame separation of each segment. Then to disperse the transmitted energy more uniformly, to reduce patterning and to break up long duration 'burst' errors, the whole 704-bit word is scrambled, and subsequently unscrambled, as noted above, according to a programme determined by a pseudo-random number generator.

On reception, the frame alignment and control information words are interrogated to identify and isolate the frame, to unscramble and de-interlace the 704-bit input signal, to determine whether the signal contains sound or data, and to check whether the signal is the same as that of the normal TV broadcast – which would allow the analogue channel to take over in the event of failure of the digital signal. If the data group identifies the input as an audio signal the 3-bit scale factor instruction is used to restore the input 10-bit amplitude range to its original 14-bit resolution.

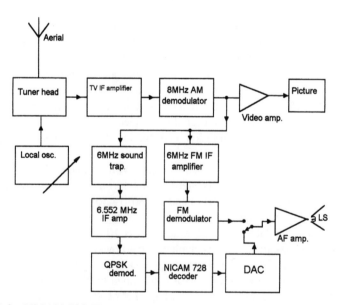

Fig. 11.6 *NICAM 728 TV receiver*

There is a time delay of about 13 ms needed for the temporary storage of the input signal during this interrogation and scale factor adjustment process. There is, in addition, a further degradation of 3 dB in the s/n ratio, which is not thought to be important for any single stage. The digital output from the NICAM demultiplexer stage is taken to a normal d/a converter, filter and audio amplifier, as shown in the schematic diagram of Fig. 11.6.

Obviously, the complexity of the NICAM process would make the construction of this type of system too costly for it to be assembled economically from basic microprocessor and RAM/ROM components, so some cross-licensing arrangements will, in due course, allow the provision of suitable application specific ICs for this purpose. Meanwhile, although there are a number of NICAM based TVs on sale, the greatest market penetration at the time of writing (1998) has been in the field of video recorders.

SOUND QUALITY

The reports of the hi-fi buffs on the quality of the output sound signal from the NICAM system are that the sound quality is good when the system is used to reproduce relatively large scale musical ensembles at high levels, but that the granularity and high quantisation noise levels associated with the basic 10-bit encoding becomes more noticeable when the system is used to reproduce single instruments or voices at low signal levels. Since the details of the sound system to be used with the projected digital TV are not yet available, it is not, at the moment, clear to what extent this will undercut NICAM-728 as the TV sound system of choice. Certainly, the facility of offering different sound channels for the same picture – for use in multi-lingual communities – is a substantial advantage offered by the NICAM system.

FURTHER READING

Bower, A.J., Digital radio *Proc. Nav. Pos. Conf.* '98. Vol. 2, pp. 40–51
Pohlmann, K.C., *Principles of Digital Audio*, 3rd edition, McGraw-Hill, ISBN 0-07-050468-7

Index